PATTERN CLASSIFICATION
USING
ENSEMBLE METHODS

SERIES IN MACHINE PERCEPTION AND ARTIFICIAL INTELLIGENCE*

Editors: **H. Bunke** (Univ. Bern, Switzerland)
P. S. P. Wang (Northeastern Univ., USA)

*For the complete list of titles in this series, please write to the Publisher.

Series in Machine Perception and Artificial Intelligence – Vol. 75

PATTERN CLASSIFICATION
USING
ENSEMBLE METHODS

Lior Rokach

Ben-Gurion University of the Negev, Israel

World Scientific

NEW JERSEY · LONDON · SINGAPORE · BEIJING · SHANGHAI · HONG KONG · TAIPEI · CHENNAI

Published by

World Scientific Publishing Co. Pte. Ltd.

5 Toh Tuck Link, Singapore 596224

USA office: 27 Warren Street, Suite 401-402, Hackensack, NJ 07601

UK office: 57 Shelton Street, Covent Garden, London WC2H 9HE

British Library Cataloguing-in-Publication Data
A catalogue record for this book is available from the British Library.

Series in Machine Perception and Artificial Intelligence — Vol. 75
PATTERN CLASSIFICATION USING ENSEMBLE METHODS

Copyright © 2010 by World Scientific Publishing Co. Pte. Ltd.

ISBN-13 978-981-4271-06-6
ISBN-10 981-4271-06-3

Printed in Singapore.

To my wife, Ronit and
my three boys, Yarden, Roy and
Amit who was born the same day this book was completed
–L.R.

Preface

Ensemble methodology imitates our second nature to seek several opinions before making a crucial decision. The core principle is to weigh several individual pattern classifiers, and combine them in order to reach a classification that is better than the one obtained by each of them separately.

Researchers from various disciplines such as pattern recognition, statistics, and machine learning have explored the use of ensemble methods since the late seventies. Given the growing interest in the field, it is not surprising that researchers and practitioners have a wide variety of methods at their disposal. Pattern Classification Using Ensemble Methods aims to provide a methodic and well structured introduction into this world by presenting a coherent and unified repository of ensemble methods, theories, trends, challenges and applications.

Its informative, factual pages will provide researchers, students and practitioners in industry with a comprehensive, yet concise and convenient reference source to ensemble methods. The book describes in detail the classical methods, as well as extensions and novel approaches that were recently introduced. Along with algorithmic descriptions of each method, the reader is provided with a description of the settings in which this method is applicable and with the consequences and the trade-offs incurred by using the method. This book is dedicated entirely to the field of ensemble methods and covers all aspects of this important and fascinating methodology.

The book consists of seven chapters. Chapter 1 presents the pattern recognition foundations that are required for reading the book. Chapter 2 introduces the basic algorithmic framework for building an ensemble of classifiers. Chapters 3-6 present specific building blocks for designing and implementing ensemble methods. Finally, Chapter 7 discusses how ensembles should be evaluated. Several selection criteria are proposed - all are presented from a practitioner's standpoint - for choosing the most effective ensemble method.

Throughout the book, special emphasis was put on the extensive use of illustrative examples. Accordingly, in addition to ensemble theory, the reader is also provided with an abundance of artificial as well as real-world applications from a wide range of fields. The data referred to in this book, as well as most of the Java implementations of the presented algorithms, can be obtained via the Web.

One of the key goals of this book is to provide researchers in the fields of pattern recognition, information systems, computer science, statistics and management with a vital source of ensemble techniques. In addition, the book will prove to be highly beneficial to those engaged in research in social sciences, psychology, medicine, genetics, and other fields that confront complex data-processing problems.

The material in this book constitutes the core of undergraduate and graduates courses at Ben-Gurion University. The book can also serve as an excellent reference book for graduate as well as advanced undergraduate courses in pattern recognition, machine learning and data mining. Descriptions of real-world data-mining projects that utilize ensemble methods may be of particular interest to the practitioners among the readers. The book is rigorous and requires comprehension of problems and solutions via their mathematical descriptions. Nevertheless, only basic background knowledge of basic probability theory and computer science (algorithms) in assumed in most of the book.

Due to the broad spectrum of ensemble methods, it is impossible to cover all techniques and algorithms in a single book. The interested reader can refer to the excellent book "pattern classifiers: methods and algorithms" by Ludmila Kuncheva (John Wiley & Sons, 2004). Other key sources include journals and conferences' proceedings. The Information Fusion Journal and the Journal of Advances in Information Fusion are largely dedicated to the field of ensemble methodology. Nevertheless, many pattern recognition, machine learning and data mining journals include research papers on ensemble techniques. Moreover, major conferences such as the International Workshop on Multiple Classifier Systems (MCS) and the International Conference on Information Fusion (FUSION) are especially recommended as sources for additional information.

Many colleagues generously gave comments on drafts or counsel otherwise. Dr. Alon Schclar deserves special mention for his particularly detailed and insightful comments. I am indebted to Prof. Oded Maimon for lending his insight to this book. Thanks also to Prof. Horst Bunke and Prof. Patrick Shen-Pei Wang for including my book in their important series in machine perception and artificial intelligence. The author would also like to thank Mr. Steven Patt, Editor, and staff members of World Scientific Publishing for their kind cooperation throughout the writing process of this book.

Last, but certainly not least, I owe my special gratitude to my family and friends for their patience, time, support, and encouragement.

Lior Rokach
Ben-Gurion University of the Negev
Beer-Sheva, Israel
September 2009

Contents

Chapter 1

Introduction to Pattern Classification

Pattern recognition is the scientific discipline whose purpose is to classify patterns (also known as instances, tuples and examples) into a set of categories which are also referred to as *classes* or *labels*. Commonly, the classification is based on statistical models that are induced from an exemplary set of preclassified patterns. Alternatively, the classification utilizes knowledge that is supplied by an expert in the application domain.

A pattern is usually composed of a set of measurements that characterize a certain object. For example, suppose we wish to classify flowers from the Iris genus into their subgeni (such as Iris Setosa, Iris Versicolour and Iris Virginica). The patterns in this case will consist the flowers features, such as the length and the width of the sepal and petal. The label of every instance will be one of the strings *Iris Setosa*, *Iris Versicolour* and *Iris Virginica*. Alternatively, the labels can take a value from 1,2,3, a,b,c or any other set of three distinct values.

Another common application which employs pattern recognition is Optical character recognition (OCR). These applications convert scanned documents into machine-editable text in order to simplify their storage and retrieval. Each document undergoes three steps. First, an operator scans the document. This converts the document into a bitmap image. Next, the scanned document is segmented such that each character is isolated from the others. Then, a *feature extractor* measures certain features of each character such as open areas, closed shapes, diagonal lines and line intersections. Finally, the scanned characters are associated with their corresponding alpha-numeric character. The association is obtained by applying a pattern recognition algorithm to the features of the scanned characters. In this case, the set of labels/categories/classes are the set of alpha-numeric character i.e. letters, numbers, punctuation marks, etc.

1.1 Pattern Classification

In a typical statistical pattern recognition setting, a set of patterns S, also referred to as a *training set* is given. The labels of the patterns in S are known and the goal is to construct an algorithm in order to label new patterns. A classification algorithm is also known as an *inducer* and an instance of an inducer for a specific training set is called a *classifier*.

The training set can be described in a variety of ways. Most frequently, each pattern is described by a vector of *feature* values. Each vector belongs to a single class and associated with the class label. In this case, the training set is stored in a table where which each row consists of a different pattern. Let A and y denote the set of n features: $A = \{a_1, \ldots, a_i, \ldots, a_n\}$ and the class label, respectively.

Features, which are also referred to as attributes, typically fall into one of the following two categories:

Nominal the values are members of an unordered set. In this case, it is useful to denote its domain values by $dom(a_i) = \{v_{i,1}, v_{i,2}, \ldots, v_{i,|dom(a_i)|}\}$, where $|dom(a_i)|$ is the finite cardinality of the domain.

Numeric the values are real numbers. Numeric features have infinite cardinalities.

In a similar way, $dom(y) = \{c_1, c_2, \ldots, c_k\}$ constitutes the set of labels. Table 1.1 illustrates a segment of the Iris dataset. This is one of the best known datasets in the pattern recognition literature. It was first introduced by R. A. Fisher (1936). The goal in this case is to classify flowers into the Iris subgeni according to their characteristic features.

The dataset contains three classes that correspond to three types of iris flowers: $dom(y) = \{IrisSetosa, IrisVersicolor, IrisVirginica\}$. Each pattern is characterized by four numeric features (measured in centimeters): $A = \{sepallength, sepalwidth, petallength, petalwidth\}$.

The instance space (the set of all possible examples) is defined as a Cartesian product of all the input attributes domains: $X = dom(a_1) \times dom(a_2) \times \ldots \times dom(a_n)$. The universal instance space (or the *labeled instance space*) U is defined as a Cartesian product of all input attribute domains and the target attribute domain, i.e.: $U = X \times dom(y)$.

The training set is denoted by $S(B)$ and it is composed of m tuples.

Table 1.1 The Iris Dataset Consisting of Four Numeric Features and Three Possible Classes.

Sepal Length	Sepal Width	Petal Length	Petal Width	Class (Iris Type)
5.1	3.5	1.4	0.2	Iris-setosa
4.9	3.0	1.4	0.2	Iris-setosa
6.0	2.7	5.1	1.6	Iris-versicolor
5.8	2.7	5.1	1.9	Iris-virginica
5.0	3.3	1.4	0.2	Iris-setosa
5.7	2.8	4.5	1.3	Iris-versicolor
5.1	3.8	1.6	0.2	Iris-setosa
⋮				

$$S(B) = (\langle x_1, y_1 \rangle, \ldots, \langle x_m, y_m \rangle) \tag{1.1}$$

where $x_q \in X$ and $y_q \in dom(y), q = 1, \ldots, m$.

Usually, it is assumed that the training set tuples are randomly generated and are independently distributed according to some fixed and unknown joint probability distribution D over U. Note that this is a generalization of the deterministic case in which a supervisor classifies a tuple using a function $y = f(x)$.

As mentioned above, the goal of an inducer is to generate classifiers. In general, a classifier partitions the instance space according to the labels of the patterns in the training set. The borders separating the regions are called *frontiers* and inducers construct frontiers such that new patterns will be classified into the correct region. Specifically, given a training set S with input attributes set $A = \{a_1, a_2, \ldots, a_n\}$ and a nominal target attribute y from an unknown fixed distribution D as defined above, the objective is to induce an optimal classifier with a minimum generalization error.

The generalization error is defined as the misclassification rate over the distribution D. Formally, let I be an inducer. We denote by $I(S)$ the classifier that is generated by I for the training set S. The classification that is produced by $I(S)$ when it is applied to a pattern x is denoted by $I(S)(x)$. In case of nominal attributes, the generalization error be expressed as:

$$\varepsilon(I(S), D) = \sum_{\langle x, y \rangle \in U} D(x, y) \cdot L(y, I(S)(x)) \tag{1.2}$$

where $L(y, I(S)(x))$ is the zero one loss function defined as:

$$L\left(y, I\left(S\right)\left(x\right)\right) = \begin{cases} 0 \; if \, y = I\left(S\right)\left(x\right) \\ 1 \; if \, y \neq I\left(S\right)\left(x\right) \end{cases} \tag{1.3}$$

In case of numeric attributes the sum operator is replaced with the integration operator.

1.2   Induction Algorithms

An *induction algorithm*, or more concisely an *inducer* (also known as learner), is an algorithm that is given a training set and constructs a model that generalizes the connection between the input attributes and the target attribute. For example, an inducer may take as input specific training patterns with their corresponding class labels, and produce a *classifier*.

Let I be an inducer. We denote by $I\left(S\right)$ the classifier which is induced by applying I to the training set S. Using $I\left(S\right)$ it is possible to predict the target value of a pattern x. This prediction is denoted as $I\left(S\right)\left(x\right)$.

Given the on going fruitful research and recent advances in the field of pattern classification, it is not surprising to find several mature approaches to induction that are now available to the practitioner.

An essential component of most classifiers is model which specifies how a new pattern is classified. These models are represented differently by different inducers. For example, C4.5 [Quinlan (1993)] represents the model as a decision tree while the Naïve Bayes [Duda and Hart (1973)] inducer represents a model in the form of probabilistic summaries. Furthermore, inducers can be deterministic (as in the case of C4.5) or stochastic (as in the case of back propagation)

There two ways in which a new pattern can be classified. The classifier can either explicitly assign a certain class to the pattern (Crisp Classifier) or, alternatively, the classifier can produce a vector of the conditional probability the given pattern belongs to each class (Probabilistic Classifier). In this case it is possible to estimate the conditional probability $\hat{P}_{I(S)}\left(y = c_j \, |a_i = x_{q,i} \; ; i = 1, \ldots, n\right)$ for an observation x_q. Note the addition of the "hat" to the conditional probability estimation is used for distinguishing it from the actual conditional probability. Inducers that can construct Probabilistic Classifiers are known as *Probabilistic Inducers*.

The following sections briefly review some of the common approaches to concept learning: Decision tree induction, Neural Networks, Genetic

Algorithms, instance-based learning, statistical methods, Bayesian methods and Support Vector Machines. This review focuses on methods that are described in details in this book.

1.3 Rule Induction

Rule induction algorithms generate a set of *if-then* rules that describes the classification process. The main advantage of this approach is its high comprehensibility. Namely, the rules can be written as a collection of consecutive conditional statements in plain English which are easy to employ. Most of the Rule induction algorithms are based on the separate and conquer paradigm [Michalski (1983)]. Consequently, these algorithms: (a) are capable of finding simple axis parallel frontiers; (b) well suited for symbolic domains; and (c) can often dispose easily of irrelevant attributes. However, Rule induction algorithms can have difficulty when non-axis-parallel frontiers are required to correctly classify the data. Furthermore, they suffer from the *fragmentation problem*, i.e., the available data dwindles as induction progresses [Pagallo and Huassler (1990)]. Another pitfall that should be avoided is the small disjuncts problem or emphoverfitting. This problem is characterized by rules that cover a very small number of training patterns. Thus, the model fits the training data very well, however, it fails to classify new patterns, resulting in a high error rate [Holte *et al.* (1989)].

1.4 Decision Trees

A Decision tree is a classifier whose model forms a recursive partition of the instance space. The model is described as a rooted tree, i.e., a direct tree with a node called a "root" that has no incoming edges. All other nodes have exactly one incoming edge. A node with outgoing edges is referred to as an "internal" node or a "test" nodes. All other nodes are called "leaves" (also known as "terminal" nodes or "decision" nodes). In a decision tree, each internal node splits the instance space into two or more sub-spaces according to a certain discrete function of the input attribute values. In the simplest and most frequent case, each test considers a single attribute, such that the instance space is partitioned according to the attributes value. In the case of numeric attributes, the condition refers to a range.

Each leaf is assigned to one class corresponding to the most appropriate target value. Alternatively, the leaf may hold a probability vector (affinity vector) whose elements indicate the probabilities of the target attribute to assume a certain value. Figure 1.1 describes an example of a decision tree that solves the Iris recognition task presented in Table 1.1.

Internal nodes are represented as circles, whereas leaves are denoted by triangles. Each internal node (not a leaf) may have two or more outgoing branches. Each node corresponds to a certain property and the branches correspond to a range of values. These value ranges must partition the set of values of the given property.

Instances are classified by traversing the tree starting from the root down to a leaf where the path is determined according to the outcome of the partitioning condition at each node. Specifically, we start at the root of a tree and consider the property that corresponds to the root. We then find to which outgoing branch the observed value of the given property corresponds. The next node in our path is the one at the end of the chosen branch. We repeat the same operations for this node and traverse the tree until we reach a leaf.

Note that decision trees can incorporate both nominal and numeric attributes. In case of numeric attributes, decision trees can be geometrically interpreted as a collection of hyperplanes, each orthogonal to one of the axes.

Naturally, decision makers prefer a less complex decision tree, as it is considered more comprehensible. Furthermore, according to [Breiman *et al.* (1984)] the tree complexity has a crucial effect on its performance accuracy. Usually, large trees are obtained by over fitting the data and hence exhibit poor generalization ability (a pitfall they share with Rule Classifiers). Nevertheless, a large decision tree can generalize well to new patterns if it was induced without over fitting the data. The tree complexity is explicitly controlled by the stopping criteria used for the construction of the tree and the pruning method that is employed. Common measures for the tree complexity include the following metrics: (i) The total number of nodes; (ii) Total number of leaves; (iii) Tree Depth; and (iv) The number of attributes used.

Decision tree induction is closely related to rule induction. Each path from the root of a decision tree to one of its leaves can be transformed into a rule simply by conjoining the tests along the path to form the antecedent part, and taking the leaf's class prediction as the class value. The resulting rule set can then be simplified in order to improve its comprehensibility to a human user, and to improve its accuracy [Quinlan (1987)].

Decision tree inducers are algorithms that automatically construct a

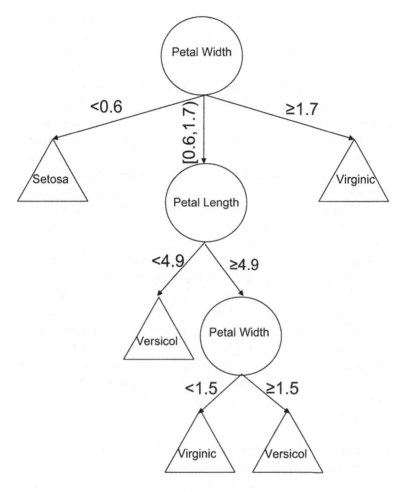

Fig. 1.1 Decision tree for solving the Iris classification task.

decision tree from a given dataset. Typically, the goal is to find the optimal decision tree by minimizing the generalization error. However, other target functions can also be defined, for instance, minimizing the number of nodes or minimizing the average depth of the tree.

Construction of an *optimal* decision tree based on a given dataset is considered to be a difficult task. [Hancock *et al.* (1996)] have shown that finding a *minimal* decision tree for a given training set is NP-hard while [Hyafil and Rivest (1976)] have proved that constructing a minimal binary tree with respect to the expected number of tests required for classifying an unseen instance is NP-complete. Even finding the minimal equivalent decision tree for a given decision tree [Zantema and Bodlaender (2000)] or building the optimal decision tree from decision tables is known to be NP-hard [Naumov (1991)].

These results indicate that optimal decision tree algorithms are only suitable for very small datasets and a very small number of attributes. Consequently, heuristic methods are required for solving this problem. Roughly speaking, these methods can be divided into two groups: methods that employ a top-down approach and methods that follow a bottom-up methodology with clear preference in the literature to the first group.

There are various top-down decision trees inducers such as ID3 [Quinlan (1986)], C4.5 [Quinlan (1993)], CART [Breiman *et al.* (1984)]. Some inducers consist of two conceptual phases: *Growing* and *Pruning* (C4.5 and CART). Other inducers perform only the growing phase.

Figure 1.2 shows an example of typical pseudo code for a top-down inducing algorithm of a decision tree using growing and pruning. Note that these algorithms are greedy by nature and construct the decision tree in a top-down, recursive manner (also known as divide and conquer). In each iteration, the algorithm considers the partition of the training set using the outcome of discrete input attributes. The selection of the most appropriate attribute is made according to a splitting measure. After an appropriate split is selected, each node further divides the training set into smaller subsets, until a stopping criterion is met.

Common stopping rules include:

(1) All instances in the training set belong to a single value of y.
(2) A maximum tree depth has been reached.
(3) The number of cases in the terminal node is less than the minimum number of cases for parent nodes.
(4) In case a node is split: the number of cases in one or more child nodes is less than the minimum number of cases for child nodes.
(5) The best splitting criterion is not greater than a certain threshold.

1.5 Bayesian Methods

1.5.1 *Overview*

Bayesian approaches employ probabilistic concept modeling, and range from the Naïve Bayes [Domingos and Pazzani (1997)] to Bayesian networks. The basic assumption of Bayesian reasoning is that the attributes are connected via a probability Moreover, when the problem at hand is supervised, the objective is to find the conditional distribution of the target attribute given the input attribute.

1.5.2 *Naïve Bayes*

1.5.2.1 *The Basic Naïve Bayes Classifier*

The most straightforward Bayesian learning method is the Naïve Bayesian inducer [Duda and Hart (1973)]. This method uses a set of discriminant functions for estimating the probability a given instance belongs to a certain class. Specifically, given an instance, it uses Bayes rule to compute the probability of each possible value of the target attribute, assuming the input attributes are conditionally independent.

Due to the fact that this method is based on the simplistic, and rather unrealistic, assumption that the causes are conditionally independent given the effect, this method is well known as Naïve Bayes.

TreeGrowing $(S, A, y, SplitCriterion, StoppingCriterion$
Where:
S - Training Set
A - Input Feature Set
y - Target Feature
$SplitCriterion$ - the method for evaluating a certain split
$StoppingCriterion$ - the criteria to stop the growing process

Create a new tree T with a single root node.
IF $StoppingCriterion(S)$ THEN
 Mark T as a leaf with the most
 common value of y in S as a label.
ELSE
 $\forall a_i \in A$ find a that obtain the best $SplitCriterion(a_i, S)$.
 Label t by a
 FOR each outcome v_i of a:
  Set $Subtree_i$= TreeGrowing $(\sigma_{a=v_i} S, A, y)$.
  Connect the root node of t_T to $Subtree_i$ with
   an edge that is labeled as v_i
 END FOR
END IF
RETURN TreePruning (S, T, y)

TreePruning (S, T, y)
Where:
S - Training Set
y - Target Feature
T - The tree to be pruned
DO
 Select a node t in T such that pruning it
  maximally improve some evaluation criteria
 IF $t \neq \emptyset$ THEN $T = pruned(T, t)$
UNTIL $t = \emptyset$
RETURN T

Fig. 1.2 Top-down algorithmic framework for decision trees induction.

The class of the instance is determined according to value of the target attribute which maximizes the following calculated probability:

$$v_{MAP}(x_q) = \underset{c_j \in dom(y)}{\mathrm{argmax}} \; \hat{P}(y = c_j) \cdot \prod_{i=1}^{n} \hat{P}(a_i = x_{q,i} \mid y = c_j) \qquad (1.4)$$

where $\hat{P}(y = c_j)$ denotes the estimation of the *a-priori* probability of the target attribute to obtain the value c_j. Similarly, $\hat{P}(a_i = x_{q,i} \mid y = c_j)$ denotes the conditional probability of the input attribute a_i to obtain the value $x_{q,i}$ given that the target attribute obtains the value c_j. Note that the hat above the conditional probability distinguishes the probability estimation from the actual conditional probability.

A simple estimation for the above probabilities can be obtained using the corresponding frequencies in the training set, namely:

$$\hat{P}(y = c_j) = \frac{|\sigma_{y=c_j} S|}{|S|} \quad ; \quad \hat{P}(a_i = x_{q,i} \mid y = c_j) = \frac{|\sigma_{y=c_j \; AND \; a_i = x_{q,i}} S|}{|\sigma_{y=c_j} S|}$$

where $|\sigma_{y=c_j} S|$ denotes the number of instances in S for which $y = c_j$. Using the Bayes rule, the above equations can be rewritten as:

$$v_{MAP}(x_q) = \underset{c_j \in dom(y)}{\mathrm{argmax}} \; \frac{\prod_{i=1}^{n} \hat{P}(y=c_j \mid a_i = x_{q,i})}{\hat{P}(y=c_j)^{n-1}} \qquad (1.5)$$

Or alternatively, after applying the log function as:

$$v_{MAP}(x_q) = \underset{c_j \in dom(y)}{\mathrm{argmax}} \; \log\left(\hat{P}(y = c_j)\right)$$
$$+ \sum_{i=1}^{n} \left(\log\left(\hat{P}(y = c_j \mid a_i = x_{q,i})\right) - \log\left(\hat{P}(y = c_j)\right) \right)$$

If the "naive" assumption is true, by a direct application of Bayes' Theorem, this classifier can easily be shown to be optimal (i.e. minimizing the generalization error), in the sense of minimizing the misclassification rate or zero-one loss (misclassification rate). It was shown in [Domingos and Pazzani (1997)] that the Naïve Bayes inducer can be optimal under zero-one loss even when the independence assumption is violated by a wide margin. This implies that the Bayesian classifier has a much greater range of applicability than originally assumed, for instance for learning conjunctions and disjunctions. Moreover, numerous empirical results show surprisingly that this method can perform quite well compared to other methods, even in domains where clear attribute dependencies exist.

The computational complexity of Naïve Bayes is considered very low compared to other methods like decision trees, since no explicit enumeration of possible interactions of various causes is required. More specifically since the Naïve Bayesian classifier combines simple functions of univariate densities, the complexity of this procedure is $O(nm)$. Furthermore, Naïve Bayes classifiers are also very simple and easy to understand [Kononenko (1990)]. Other advantages of Naïve Bayes include easy adaptation of the model to incremental learning environments and robustness to irrelevant attributes. The main disadvantage of the Naïve Bayes inducer is that it is limited to only simplified models, which in some cases are incapable of representing the complicated nature of the problem. To understand this weakness, consider a target attribute that cannot be explained by a single attribute, for instance, the Boolean exclusive or function (XOR).

The Naïve Bayesian classifier uses all the available attributes, unless a feature selection procedure is applied as a pre-processing step.

1.5.2.2 *Naïve Bayes Induction for Numeric Attributes*

Originally, Naïve Bayes assumes that all input attributes are nominal. If this is not the case then there are some options to bypass this problem:

(1) Pre-Processing: The numeric attributes are discretized before using the Naïve Bayes approach. It is suggested in [Domingos and Pazzani (1997)] to construct ten equi-length intervals for each numeric attribute (or one per observed value, whichever produces the least number of possible values). Each attribute value will be assigned an interval number. Obviously, there are many other more context-aware discretization methods that can be applied here and probably obtain better results.
(2) Revising the Naïve Bayes: [John and Langley (1995)] suggests using kernel estimation or single variable normal distribution as part of the conditional probabilities calculation.

1.5.2.3 *Correction to the Probability Estimation*

Using the probability estimation described above as-is will typically over-estimate (similar to over-fitting in decision trees) the probability. This can be problematic especially when a given class and attribute value never co-occur in the training set. This case leads to a zero probability that wipes out the information in all the other probabilities terms when they are multiplied according to the original Naïve Bayes equation.

There are two known corrections for the simple probability estimation which circumvent this phenomenon. The following sections describe these corrections.

1.5.2.4 *Laplace Correction*

According to Laplace's law of succession [Niblett (1987)], the probability of the event $y = c_i$ (y is a random variable and c_i is a possible outcome of y) which is observed m_i times out of m observations is:

$$\frac{m_i + k p_a}{m + k}$$

where p_a is an *a-priori* probability estimation of the event and k is the equivalent sample size that determines the weight of the *a-priori* estimation relative to the observed data. According to [Mitchell (1997)], k is called "equivalent sample size" because it represents an augmentation of the m actual observations by additional k virtual samples distributed according to p_a. The above ratio can be rewritten as the weighted average of the *a-priori* probability and the posteriori probability (denoted as p_p):

$$\begin{aligned}
&\frac{m_i + k \cdot p_a}{m + k} \\
&= \frac{m_i}{m} \cdot \frac{m}{m+k} + p_a \cdot \frac{k}{m+k} \\
&= p_p \cdot \frac{m}{m+k} + p_a \cdot \frac{k}{m+k} = \\
&= p_p \cdot w_1 + p_a \cdot w_2
\end{aligned}$$

In the case discussed here the following correction is used:

$$\hat{P}_{Laplace}\left(a_i = x_{q,i} \mid y = c_j\right) = \frac{\left|\sigma_{y=c_j \ AND \ a_i=x_{q,i}} S\right| + k \cdot p}{\left|\sigma_{y=c_j} S\right| + k} \tag{1.6}$$

In order to use the above correction, the values of p and k should be determined. There are several possibilities to determine their values. It is possible to use $p = 1/\left|dom\left(y\right)\right|$ and $k = \left|dom\left(y\right)\right|$. In [Ali and Pazzani (1996)] it is suggested to use $k = 2$ and $p = 1/2$ in any case even if $\left|dom\left(y\right)\right| > 2$ in order to emphasize the fact that the estimated event is always compared to the opposite event. Another option is to use $k = \left|dom\left(y\right)\right| / \left|S\right|$ and $p = 1/\left|dom\left(y\right)\right|$ [Kohavi *et al.* (1997)].

1.5.2.5 *No Match*

According to [Clark and Niblett (1989)] only zero probabilities should be corrected and replaced by the following value: $p_a/|S|$ where [Kohavi et al. (1997)] suggest to use $p_a = 0.5$. An empirical comparison of Laplace correction and No-Match correction indicates that there is no significant difference between them. However, both of them are significantly better than not performing any correction at all.

1.5.3 *Other Bayesian Methods*

A more sophisticated Bayesian-based model that can be used is Bayesian belief networks [Pearl (1988)]. Usually each node in a Bayesian network represents a certain attribute. The immediate predecessors of a node represent the attributes on which the node depends. By knowing their values, it is possible to determine the conditional distribution of this node. Bayesian networks have the benefit of a clearer semantics than more ad hoc methods, and they provide a natural platform for combining domain knowledge (in the initial network structure) and empirical learning (of the probabilities, and possibly of a new structure). However, time complexity of inference in Bayesian networks can be high, and as tools for classification learning they are not yet as mature or well tested as other approaches. More generally, as [Buntine (1990)] notes, the Bayesian paradigm extends beyond any single representation, and forms a framework in which many learning tasks can be usefully studied.

1.6   Other Induction Methods

1.6.1 *Neural Networks*

Neural network methods construct a model using a network of interconnected units called neurons[Anderson and Rosenfeld (2000)]. The neurons are connected in an input/output manner i.e. the output of one neuron (antecedent) is the input of another (descendant). A neuron may have several antecedents and several descendants (including itself in some settings). Every unit performs a simple data processing task by generating an output from the received inputs. The task is usually obtained via a nonlinear function. The most frequently used type of unit, incorporating Sigmoidal nonlinearity, can be seen as a generalization of a propositional rule, where numeric weights are assigned to antecedents, and the output is graded, rather than binary [Towell and Shavlik (1994)].

The multilayer feedforward neural network is the most widely studied neural network, because it is suitable for representing functional relationships between a set of input attributes and one or more target attributes. In a multilayer feedforward neural network the neurons are organized in layers. Figure 1.3 illustrates a typical feedforward neural network. This network consists of neurons (also referred to as nodes) organized in three layers: an input layer, a hidden layer and an output layer. The neurons in the input layer correspond to the input attributes and the neurons in the output layer correspond to the target attribute. The neurons in the hidden layer are connected to both the input and the output neurons and they are the key to the induction of the classifier. Note that the signal flow is directed from the input layer to the output layer and there are no loops.

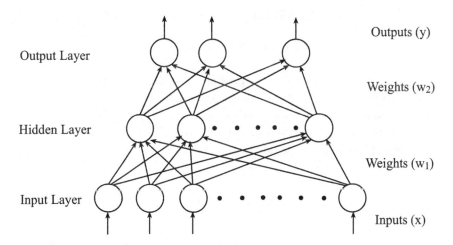

Fig. 1.3 Three-layer feedforward neural network.

In order to construct a classifier from a neural network inducer, a *training* step must be employed. The training step calculates the connection weights which optimize a given evaluation function of the training data. Various search methods can be used to train these networks, of which the most widely applied one is back propagation [Rumelhart *et al.* (1986)]. This method efficiently propagates values of the output evaluation function backward to the input, allowing the network weights to be adapted so as to obtain a better evaluation score. Radial basis function (RBF) networks employ Gaussian nonlinearity in the neurons [Moody and Darken (1989)], and can be seen as a generalization of nearestneighbor methods with an exponential distance function [Poggio and Girosi (1990)].

Most neural networks are based on a unit called a *perceptron*. A perceptron performs the following: (a) it calculates a linear combination of its inputs; and (b) it invokes an activation function which transforms the weighted sum into a binary output. Figure 1.4 illustrates the perceptron.

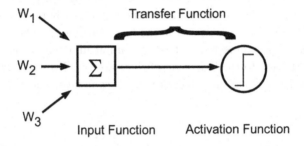

Fig. 1.4   The perceptron.

Using a single perceptron, it is possible to realize any binary decision function (two-class classification) that can be modeled as a hyper-plane in the attribute space of the input. Any instance on one side of the hyperplane is assigned to one class, and instances on the other side are assigned to the other class. The equation for this hyperplane is:

$$\sum_{i=1}^{n} w_i \cdot x_i = 0$$

where each w_i is a real-valued weight, that determines the contribution of each input signal x_i to the perceptron output.

Neural networks are remarkable for their learning efficiency and tend to outperform other methods (like decision trees) when the information required for the classification is not concentrated in a small subset of the attributes, but rather it is spread across many of the attributes. Furthermore, neural networks can be trained incrementally i.e. they can easily be adjusted as new training examples become available.

However, according to [Lu *et al.* (1996)], the drawbacks of applying neural networks to data mining include: difficulty in interpreting the model, difficulty in incorporating prior knowledge about the application domain, and, also, long training time, both in terms of CPU time, and of manually finding parameter settings that will enable successful learning i.e. optimize the evaluation function. The rule extraction algorithm, described in [Lu *et al.* (1996)], makes an effective use of the neural network structure. The algorithm extracts the (Boolean) rules in a deterministic manner without using the connection weights. The network is then pruned by removal of redundant links and units with the exception of attributes (Feature selection) whose removal is not considered.

1.6.2 *Genetic Algorithms*

Genetic algorithms are a collection of search methods that can be used to train a wide variety of models. - of which the most frequently used one is probably *rule sets* [Booker *et al.* (1989)]. Genetic algorithms maintain a population of classifiers during the training, as opposed to just one in other search methods. They employ an iterative process whose goal is to find an optimal classifier by improving the evaluation performance of the classifier population. In order to achieve this, pairs of classifiers that achieve better performance are chosen. Random mutations are plied to the classifier pair and part are exchanged between them. This process has a lower chance to reach a local minima than simple greedy search employed in most learners do. However, this process may incur a high computational cost. Furthermore, there is a higher risk of producing poor classifiers that accidently perform well on the training data.

1.6.3 *Instance-based Learning*

Instance-based learning algorithms [Aha *et al.* (1991)] are non-parametric general classification algorithms that classify a new unlabeled instance according to the labels of similar instances in the training set. At the core of these algorithms, there is a simple search procedure. These techniques are able to induce complex classifiers from a relatively small number of examples and are naturally suited to numeric domains. However, they can be very sensitive to irrelevant attributes and are unable to select different attributes in different regions of the instance space. Furthermore, although (or more accurately *because*) the time complexity to train these models is low, it is relatively time consuming to classify a new instance.

The most basic and simplest Instance-based method is the nearest neighbor (NN) inducer, which was first examined by [Fix and Hodges (1957)]. It can be represented by the following rule: to classify an unknown pattern, choose the class of the nearest example in the training set as measured by a given distance metric. A common extension is to choose the most common class in the *k nearest neighbors* (kNN).

Despite its simplicity, the nearest neighbor classifier has many advantages over other methods. For instance, it can generalize from a relatively small training set. Namely, compared to other methods, such as decision trees or neural network, the nearest neighbor classifier requires smaller training examples to achieve the same classification performance. Moreover, new information can be incrementally incorporated at runtime a property it shares with neural networks. consequently, the nearest neighbor classifier can achieve a performance that is competitive to more modern and complex methods such as decision trees or neural networks.

1.6.4   *Support Vector Machines*

Support Vector Machines [Vapnik (1995)] map the input space into a high-dimensional feature space through a non-linear mapping that is chosen *a-priori*. An optimal separating hyperplane is then constructed in the new feature space. The method searches for a hyperplane that is optimal according the VC-Dimension theory.

Chapter 2

Introduction to Ensemble Learning

The main idea behind the ensemble methodology is to weigh several individual pattern classifiers, and combine them in order to obtain a classifier that outperforms every one of them. Ensemble methodology imitates our second nature to seek several opinions before making any crucial decision. We weigh the individual opinions, and combine them to reach our final decision [Polikar (2006)].

In the literature, the term "ensemble methods" is usually reserved for collections of models that are minor variants of the same basic model. Nevertheless, in this book we also cover hybridization of models that are not from the same family. The latter is also referred in the literature as "multiple classifier systems.

Successful application of ensemble methods can be found in many fields, such as: finance [Leigh *et al.* (2002)], bioinformatics [Tan *et al.* (2003)], medicine [Mangiameli *et al.* (2004)], cheminformatics [Merkwirth *et al.* (2004)], manufacturing [Maimon and Rokach (2004)], geography [Bruzzone *et al.* (2004)], and Image Retrieval [Lin *et al.* (2006)].

The idea of building a predictive model that integrates multiple models has been investigated for a long time. The history of ensemble methods dates back to as early as 1977 with Tukeys Twicing [Tukey (1977)]: an ensemble of two linear regression models. Tukey suggested to fit the first linear regression model to the original data and the second linear model to the residuals. Two years later, Dasarathy and Sheela (1979) suggested to partition the input space using two or more classifiers. The main progress in the field was achieved during the Nineties. Hansen and Salamon (1990) suggested an ensemble of similarly configured neural networks to improve the predictive performance of a single one. At the same time Schapire (1990) laid the foundations for the award winning AdaBoost [Freund and Schapire

(1996)] algorithm by showing that a strong classifier in the *probably approx- imately correct* (PAC) sense can be generated by combining "weak" clas- sifiers (that is, simple classifiers whose classification performance is only slightly better than random classification). Ensemble methods can also be used for improving the quality and robustness of unsupervised tasks. Nevertheless, in this book we focus on classifier ensembles.

In the past few years, experimental studies conducted by the machine- learning community show that combining the outputs of multiple classi- fiers reduces the generalization error [Domingos (1996); Quinlan (1996); Bauer and Kohavi (1999); Opitz and Maclin (1999)] of the individual classifiers. Ensemble methods are very effective, mainly due to the phe- nomenon that various types of classifiers have different "inductive biases" [Mitchell (1997)]. Indeed, ensemble methods can effectively make use of such diversity to reduce the variance-error [Tumer and Ghosh (1996); Ali and Pazzani (1996)] without increasing the bias-error. In certain sit- uations, an ensemble method can also reduce bias-error, as shown by the theory of large margin classifiers [Bartlett and Shawe-Taylor (1998)].

2.1   Back to the Roots

Marie Jean Antoine Nicolas de Caritat, marquis de Condorcet (1743-1794) was a French mathematician who among others wrote in 1785 the Essay on the Application of Analysis to the Probability of Majority Decisions. This work presented the well-known Condorcet's jury theorem. The theorem refers to a jury of voters who need to make a decision regarding a binary outcome (for example to convict or not a defendant). If each voter has a probability p of being correct and the probability of a majority of voters being correct is M then:

- $p > 0.5$ implies $M > p$
- Also M approaches 1, for all $p > 0.5$ as the number of voters approaches infinity.

This theorem has two major limitations: the assumption that the votes are independent; and that there are only two possible outcomes. Never- theless, if these two preconditions are met, then a correct decision can be obtained by simply combining the votes of a large enough jury that is com- posed of voters whose judgments are slightly better than a random vote.

Originally, the Condorcet Jury Theorem was written to provide a theoretical basis for democracy. Nonetheless, the same principle can be applied in pattern recognition. A strong learner is an inducer that is given a training set consisting of labeled data and produces a classifier which can be arbitrarily accurate. A weak learner produces a classifier which is only slightly more accurate than random classification. The formal definitions of weak and strong learners are beyond the scope of this book. The reader is referred to [Schapire (1990)] for these definitions under the PAC theory. A decision stump inducer is one example of a weak learner. A Decision Stump is a one-level Decision Tree with either a categorical or a numerical class label. Figure 2.1 illustrates a decision stump for the Iris dataset presented in Table 1.1. The classifier distinguished between three cases: Petal Length greater or equal to 2.45, Petal Length smaller than 2.45 and Petal Length that is unknown. For each case the classifier predict a different class distribution.

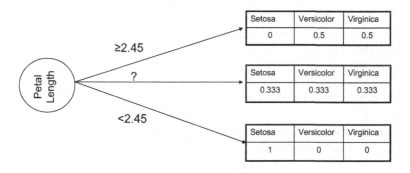

Fig. 2.1   A decision stump classifier for solving the Iris classification task.

One of the basic questions that has been investigated in ensemble learning is: "can a collection of weak classifiers create a single strong one?". Applying the Condorcet Jury Theorem insinuates that this goal might be achieved. Namely, construct an ensemble that (a) consists of independent classifiers, each of which correctly classifies a pattern with a probability of $p > 0.5$; and (b) has a probability of $M > p$ to jointly classify a pattern to its correct class.

2.2   The Wisdom of Crowds

Sir Francis Galton (1822-1911) was an English philosopher and statistician that conceived the basic concept of standard deviation and correlation. While visiting a livestock fair, Galton was intrigued by a simple weight-guessing contest. The visitors were invited to guess the weight of an ox. Hundreds of people participated in this contest, but no one succeeded to guess the exact weight: 1,198 pounds. Nevertheless, surprisingly enough, Galton found out that the average of all guesses came quite close to the exact weight: 1,197 pounds. Similarly to the Condorcet jury theorem, Galton revealed the power of combining many simplistic predictions in order to obtain an accurate prediction.

James Michael Surowiecki, an American financial journalist, published in 2004 the book "The Wisdom of Crowds: Why the Many Are Smarter Than the Few and How Collective Wisdom Shapes Business, Economies, Societies and Nations". Surowiecki argues, that under certain controlled conditions, the aggregation of information from several sources, results in decisions that are often superior to those that could have been made by any single individual - even experts.

Naturally, not all crowds are wise (for example, greedy investors of a stock market bubble). Surowiecki indicates that in order to become wise, the crowd should comply with the following criteria:

- **Diversity of opinion** – Each member should have private information even if it is just an eccentric interpretation of the known facts.
- **Independence** – Members' opinions are not determined by the opinions of those around them.
- **Decentralization** – Members are able to specialize and draw conclusions based on local knowledge.
- **Aggregation** – Some mechanism exists for turning private judgments into a collective decision.

2.3   The Bagging Algorithm

Bagging (bootstrap aggregating) is a simple yet effective method for generating an ensemble of classifiers. The ensemble classifier, which is created by this method, consolidates the outputs of various learned classifiers into a single classification. This results in a classifier whose accuracy is higher than the accuracy of each individual classifier. Specifically, each classifier in the ensemble is trained on a sample of instances taken with replacement (allowing repetitions) from the training set. All classifiers are trained using the same inducer.

To ensure that there is a sufficient number of training instances in every sample, it is common to set the size of each sample to the size of the original training set. Figure 2.2 presents the pseudo-code for building an ensemble of classifiers using the bagging algorithm [Breiman (1996a)]. The algorithm receives an induction algorithm I which is used for training all members of the ensemble. The stopping criterion in line 6 terminates the training when the ensemble size reaches T. One of the main advantages of bagging is that it can be easily implemented in a parallel mode by training the various ensemble classifiers on different processors.

Bagging Training

Require: I (a base inducer), T (number of iterations), S (the original training set), μ (the sample size).

1: $t \leftarrow 1$
2: **repeat**
3:    $S_t \leftarrow$ a sample of μ instances from S with replacement.
4:    Construct classifier M_t using I with S_t as the training set
5:    $t \leftarrow t + 1$
6: **until** $t > T$

Fig. 2.2 The Bagging algorithm.

Note that since sampling with replacement is used, some of the original instances of S may appear more than once in S_t and some may not be included at all. Furthermore, using a large sample size causes individual samples to overlap significantly, with many of the same instances appearing in most samples. So while the training sets in S_t may be different from one another, they are certainly not independent from a statistical stand point. In order to ensure diversity among the ensemble members, a relatively unstable inducer should be used. This will result is an ensemble of sufficiently different classifiers which can be acquired by applying small perturbations to the training set. If a stable inducer is used, the ensemble will be composed of a set of classifiers who produce nearly similar classifications, and thus will unlikely improve the performance accuracy.

In order to classify a new instance, each classifier returns the class prediction for the unknown instance. The composite *bagged* classifier returns the class with the highest number of predictions (also known as majority voting).

Bagging Classification

Require: x (an instance to be classified)

Ensure: C (predicted class)

1: $Counter_1, \ldots, Counter_{|dom(y)|} \leftarrow 0$ {initializes class votes counters}
2: **for** $i = 1$ to T **do**
3:    $vote_i \leftarrow M_i(x)$ {get predicted class from member i}
4:    $Counter_{vote_i} \leftarrow Counter_{vote_i} + 1$ {increase by 1 the counter of the corresponding class}
5: **end for**
6: $C \leftarrow$ the class with the largest number votes
7: Return C

Fig. 2.3   The Bagging Classification.

We demonstrate the bagging procedure by applying it to the Labor dataset presented in Table 2.1. Each instance in the table stands for a collective agreement reached in the business and personal services sectors (such as teachers and nurses) in Canada during the years 1987-1988. The aim of the learning task is to distinguish between acceptable and unacceptable agreements (as classified by experts in the field). The selected input-features that characterize the agreement are:

- Dur – the duration of agreement
- Wage – wage increase in first year of contract
- Stat – number of statutory holidays
- Vac – number of paid vacation days
- Dis – employer's help during employee longterm disability
- Dental – contribution of the employer towards a dental plan
- Ber – employer's financial contribution in the costs of bereavement
- Health – employer's contribution towards the health plan

Applying a Decision Stump inducer on the Labor dataset results in the model that is depicted in Figure 2.4. Using ten-folds cross validation, the estimated generalized accuracy of this model is 59.64%.

Next, we execute the bagging algorithm using Decision Stump as the base inducer ($I = DecisionStump$), four iterations ($T = 4$) and a sample size that is equal to the original training set ($\mu = |S| = 57$. Recall that the

Table 2.1   The Labor Dataset.

Dur	Wage	Stat	Vac	Dis	Dental	Ber	Health	Class
1	5	11	average	?	?	yes	?	good
2	4.5	11	below	?	full	?	full	good
?	?	11	generous	yes	half	yes	half	good
3	3.7	?	?	?	?	yes	?	good
3	4.5	12	average	?	half	yes	half	good
2	2	12	average	?	?	?	?	good
3	4	12	generous	yes	none	yes	half	good
3	6.9	12	below	?	?	?	?	good
2	3	11	below	yes	half	yes	?	good
1	5.7	11	generous	yes	full	?	?	good
3	3.5	13	generous	?	?	yes	full	good
2	6.4	15	?	?	full	?	?	good
2	3.5	10	below	no	half	?	half	bad
3	3.5	13	generous	?	full	yes	full	good
1	3	11	generous	?	?	?	?	good
2	4.5	11	average	?	full	yes	?	good
1	2.8	12	below	?	?	?	?	good
1	2.1	9	below	yes	half	?	none	bad
1	2	11	average	no	none	no	none	bad
2	4	15	generous	?	?	?	?	good
2	4.3	12	generous	?	full	?	full	good
2	2.5	11	below	?	?	?	?	bad
3	3.5	?	?	?	?	?	?	good
2	4.5	10	generous	?	half	?	full	good
1	6	9	generous	?	?	?	?	good
3	2	10	below	?	half	yes	full	bad
2	4.5	10	below	yes	none	?	half	good
2	3	12	generous	?	?	yes	full	good
2	5	11	below	yes	full	yes	full	good
3	2	10	average	?	?	yes	full	bad
3	4.5	11	average	?	half	?	?	good
3	3	10	below	yes	half	yes	full	bad
2	2.5	10	average	?	?	?	?	bad
2	4	10	below	no	none	?	none	bad
3	2	10	below	no	half	yes	full	bad
2	2	11	average	yes	none	yes	full	bad
1	2	11	generous	no	none	no	none	bad
1	2.8	9	below	yes	half	?	none	bad
3	2	10	average	?	?	yes	none	bad
2	4.5	12	average	yes	full	yes	half	good
1	4	11	average	no	none	no	none	bad
2	2	12	generous	yes	none	yes	full	bad
2	2.5	12	average	?	?	yes	?	bad
2	2.5	11	below	?	?	yes	?	bad
2	4	10	below	no	none	?	none	bad
2	4.5	10	below	no	half	?	half	bad
2	4.5	11	average	?	full	yes	full	good
2	4.6	?	?	yes	half	?	half	good
2	5	11	below	yes	?	?	full	good
2	5.7	11	average	yes	full	yes	full	good
2	7	11	?	yes	full	?	?	good
3	2	?	?	yes	half	yes	?	good
3	3.5	13	generous	?	?	yes	full	good
3	4	11	average	yes	full	?	full	good
3	5	11	generous	yes	?	?	full	good
3	5	12	average	?	half	yes	half	good
3	6	9	generous	yes	full	yes	full	good

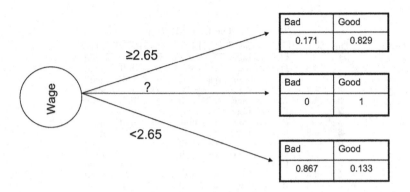

Fig. 2.4   Decision Stump classifier for solving the Labor classification task.

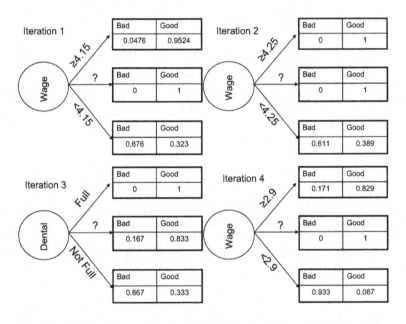

Fig. 2.5   Decision Stump classifiers for solving the Labor classification task.

sampling is performed with replacement. Consequently, in each iteration some of the original instances from S may appear more than once and some may not be included at all. Table 2.2 indicates for each instance the number of times it was sampled in every iteration.

Table 2.2 The Bagging Sample Sizes for the Labor Dataset.

Instance	Iteration 1	Iteration 2	Iteration 3	Iteration 4
1			1	1
2	1	1	1	1
3	2	1	1	1
4	1	1		
5		1	2	1
6	2	1		1
7		1	1	1
8	1		1	1
9		2		
10	1	1	2	
11		1		1
12	1	1	1	
13		1	1	2
14	1	1		2
15	1		1	1
16	2	3	2	1
17	1			
18	1	2	2	1
19	1			2
20		1	3	
21	2		1	1
22	2	1	1	
23		1	2	1
24	1	1		1
25	2	2	1	2
26	1	1	1	1
27		1		
28	2	1	1	1
29	1		1	1
30	1	1		
31		1	1	3
32	2	1	2	
33	1			1
34	2	2	1	
35				2
36	1	1	1	1
37		2	1	
38	2	1		1
39	2	1	1	
40	1	1	1	2
41	1	2	2	2
42		3	2	1
43	2			1
44	1	2	1	3
45	3	1	2	1
46	1		1	2
47	1	1	2	1
48		1	1	1
49	1	2	1	2
50	2		1	1
51	1	1	1	1
52	1	1	1	
53	1	2	2	2
54	1	1		
55			2	2
56	1	1	1	1
57	1	1	2	1
Total	57	57	57	57

Figure 2.5 presents the four classifiers that were built. The estimated generalized accuracy of the ensemble classifier that uses them rises to 77.19%

Often, bagging produces a combined model that outperforms the model that is built using a single instance of the original data. Breiman (1996) notes that this is true especially for unstable inducers since bagging can eliminate their instability. In this context, an inducer is considered unstable if perturbations in the learning set can produce significant changes in the constructed classifier.

2.4 The Boosting Algorithm

Boosting is a general method for improving the performance of a weak learner. The method works by iteratively invoking a weak learner, on training data that is taken from various distributions. Similar to bagging, the classifiers are generated by resampling the training set. The classifiers are then combined into a single strong composite classifier. Contrary to bagging, the resampling mechanism in boosting improves the sample in order to provide the most useful sample for each consecutive iteration. Breiman [Breiman (1996a)] refers to the boosting idea as the most significant development in classifier design of the Nineties.

The boosting algorithm is described in Figure 2.6. The algorithm generates three classifiers. The sample, S_1, that is used to train the first classifier, M_1, is randomly selected from the original training set. The second classifier, M_2, is trained on a sample, M_2, half of which consists of instances that are incorrectly classified by M_1 and the other half is composed of instances that are correctly classified by M_2. The last classifier M_3 is trained with instances on which the two previous classifiers disagree. In order to classify a new instance, each classifier produces its predicted class. The ensemble classifier returns the class that has the majority of the votes.

2.5 The AdaBoost Algorithm

AdaBoost (Adaptive Boosting), which was first introduced in [Freund and Schapire (1996)], is a popular ensemble algorithm that improves the simple boosting algorithm via an iterative process. The main idea behind this algorithm is to give more focus to patterns that are harder to classify. The amount of focus is quantified by a weight that is assigned to every

Boosting Training

Require: I (a weak inducer), S (training set) and k (the sample size for the first classifier)

Ensure: M_1, M_2, M_3

1: $S_1 \leftarrow$ Randomly selected $k < m$ instances from S without replacement;

2: $M_1 \leftarrow I(S_1)$

3: $S_2 \leftarrow$ Randomly selected instances (without replacement) from $S - S_1$ such that half of them are correctly classified by M_1.

4: $M_2 \leftarrow I(S_2)$

5: $S_3 \leftarrow$ any instances in $S - S_1 - S_2$ that are classified differently by M_1 and M_2.

Fig. 2.6   The Boosting algorithm.

pattern in the training set. Initially, the same weight is assigned to all the patterns. In each iteration the weights of all misclassified instances are increased while the weights of correctly classified instances are decreased. As a consequence, the weak learner is forced to focus on the difficult instances of the training set by performing additional iterations and creating more classifiers. Furthermore, a weight is assigned to every individual classifier. This weight measures the *overall accuracy* of the classifier and is a function of the total weight of the correctly classified patterns. Thus, higher weights are given to more accurate classifiers. These weights are used for the classification of new patterns.

This iterative procedure provides a series of classifiers that complement one another. In particular, it has been shown that AdaBoost approximates a large margin classifier such as the SVM [Rudin *et al.* (2004)].

The pseudo-code of the AdaBoost algorithm is described in Figure 2.7. The algorithm assumes that the training set consists of m instances, which are either labeled as -1 or $+1$. The classification of a new instance is obtained by voting on all classifiers $\{M_t\}$, each having an overall accuracy of α_t. Mathematically, it can be written as:

$$H(x) = sign(\sum_{t=1}^{T} \alpha_t \cdot M_t(x)) \qquad (2.1)$$

Breiman [Breiman (1998)] explores a simpler algorithm called Arc-x4 whose purpose it to demonstrate that AdaBoost works not because of the specific form of the weighing function, but because of the adaptive

AdaBoost Training

Require: I (a weak inducer), T (the number of iterations), S (training set)

Ensure: $M_t, \alpha_t; t = 1, \ldots, T$

1: $t \leftarrow 1$

2: $D_1(i) \leftarrow 1/m; i = 1, \ldots, m$

3: **repeat**

4:    Build Classifier M_t using I and distribution D_t

5:    $\varepsilon_t \leftarrow \sum\limits_{i: M_t(x_i) \neq y_i} D_t(i)$

6:    **if** $\varepsilon_t > 0.5$ **then**

7:       $T \leftarrow t - 1$

8:       exit Loop.

9:    **end if**

10:    $\alpha_t \leftarrow \frac{1}{2} \ln \left(\frac{1 - \varepsilon_t}{\varepsilon_t} \right)$

11:    $D_{t+1}(i) = D_t(i) \cdot e^{-\alpha_t y_t M_t(x_i)}$

12:    Normalize D_{t+1} to be a proper distribution.

13:    $t \leftarrow t + 1$

14: **until** $t > T$

Fig. 2.7   The AdaBoost algorithm.

resampling. In Arc-x4, a new pattern is classified according to unweighted voting and the updated $t + 1$ iteration probabilities are defined by:

$$D_{t+1}(i) = 1 + m_{t_i}^4 \tag{2.2}$$

where m_{t_i} is the number of misclassifications of the i-th instance by the first t classifiers.

AdaBoost assumes that the weak inducers, which are used to construct the classifiers, can handle weighted instances. For example, most decision tree algorithms can handle weighted instances. However, if this is not the case, an unweighted dataset is generated from the weighted data via resampling. Namely, instances are chosen with a probability according to their weights (until the dataset becomes as large as the original training set).

In order to demonstrate how the AdaBoost algorithm works, we apply it to the Labor dataset using Decision Stump as the base inducer. For the sake of simplicity we use a feature-reduced version of the labor dataset

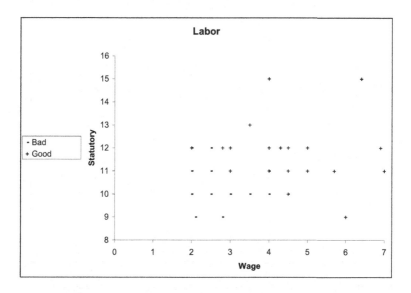

Fig. 2.8   The Labor dataset.

which is composed of two input attributes: wage and statutory. Figure 2.8 presents this projected dataset where the plus symbol represents the "good" class and the minus symbol represents the "bad" class.

The initial distribution D_1 is set to be uniform. Consequently, it is not surprising that the first classifier is identical to the decision stump that was presented in Figure 2.4. Figure 2.9 depicts the decision bound of the first classifier. The training misclassification rate of the first classifier is $\varepsilon_1 = 23.19\%$ therefore the overall accuracy weight of the first classifier is $\alpha_1 = 0.835$.

The weights of the instances are updated according to the misclassification rates as described in lines 11-12. Figure 2.10 illustrates the new weights: instances whose weights were increased are depicted by larger symbols. Table 2.3 summarizes the exact weights after every iteration.

Applying the decision stump algorithm once more produces the classifier that is shown in Figure 2.11. The training misclassification rate of the second classifier is $\varepsilon_1 = 25.94\%$, and therefore the overall accuracy weight of the second classifier is $\alpha_1 = 0.79$. The new decision boundary, which is derived from the second classifier, is illustrated in Figure 2.12.

The subsequent iterations create the additional classifiers presented in Figure 2.11. Using the ten-folds cross validation procedure, AdaBoost increased the estimated generalized accuracy from 59.64% to 87.71%, after

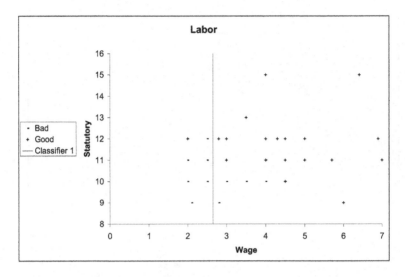

Fig. 2.9   The Labor dataset with the decision bound of the first classifier.

only four iterations. This improvement appears to be better than the improvement that is demonstrated by Bagging (77.19%). Nevertheless, AdaBoost was given a reduced version of the Labor dataset with only two input attributes. These two attributes were not selected arbitrarily. As a matter of fact, the attributes were selected based on prior knowledge so that AdaBoost will focus on the most relevant attributes. If AdaBoost is given the same dataset that was given to the Bagging algorithm, it obtains the same accuracy. Nevertheless, by increasing the ensemble size, AdaBoost continues to improve the accuracy while the Bagging algorithm shows very little improvement. For example, in case ten classifiers are trained over the full labor dataset, AdaBoost obtains an accuracy of 82.45% while Bagging obtains an accuracy of only 78.94%.

AdaBoost seems to improve the performance accuracy for two main reasons:

(1) It generates a final classifier whose error on the training set is small by combining many hypotheses whose error may be large.
(2) It produces a combined classifier whose variance is significantly lower than the variances produced by the weak base learner.

However, AdaBoost sometimes fails to improve the performance of the base inducer. According to Quinlan [Quinlan (1996)], the main reason for

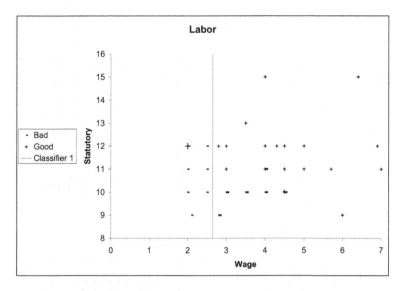

Fig. 2.10   Labor instances weights after first iteration. Instances whose weights were increased are depicted by larger symbols. The vertical line depicts the decision boundary that is derived by the first classifier.

AdaBoost's failure is overfitting. The objective of boosting is to construct a composite classifier that performs well on the data by iteratively improving the classification accuracy. Nevertheless, a large number of iterations may result in an overcomplex composite classifier, which is significantly less accurate than a single classifier. One possible way to avoid overfitting is to keep the number of iterations as small as possible.

Bagging, like boosting, is a technique that improves the accuracy of a classifier by generating a composite model that combines multiple classifiers all of which are derived from the same inducer. Both methods follow a voting approach, which is implemented differently, in order to combine the outputs of the different classifiers. In boosting, as opposed to bagging, each classifier is influenced by the performance of those that were built prior to its construction. Specifically, the new classifier pays more attention to classification errors that were done by the previously built classifiers where the amount of attention is determined according to their performance. In bagging, each instance is chosen with equal probability, while in boosting, instances are chosen with a probability that is proportional to their weight. Furthermore, as mentioned above (Quinlan, 1996), bagging requires an unstable learner as the base inducer, while in boosting inducer instability in not required, only that the error rate of every classifier be kept below 0.5.

Table 2.3   The weights of the instances in the Labor Dataset after every iteration.

				Weights		
Wage	Statutory	Class	Iteration 1	Iteration 2	Iteration 3	Iteration 4
5	11	good	1	0.59375	0.357740586	0.217005076
4.5	11	good	1	0.59375	0.357740586	0.217005076
?	11	good	1	0.59375	0.357740586	0.217005076
3.7	?	good	1	0.59375	0.357740586	1.017857143
4.5	12	good	1	0.59375	0.357740586	0.217005076
2	12	good	1	3.166666667	1.907949791	5.428571429
4	12	good	1	0.59375	0.357740586	1.017857143
6.9	12	good	1	0.59375	0.357740586	0.217005076
3	11	good	1	0.59375	0.357740586	1.017857143
5.7	11	good	1	0.59375	0.357740586	0.217005076
3.5	13	good	1	0.59375	0.357740586	1.017857143
6.4	15	good	1	0.59375	0.357740586	0.217005076
3.5	10	bad	1	3.166666667	1.907949791	1.157360406
3.5	13	good	1	0.59375	0.357740586	1.017857143
3	11	good	1	0.59375	0.357740586	1.017857143
4.5	11	good	1	0.59375	0.357740586	0.217005076
2.8	12	good	1	0.59375	0.357740586	1.017857143
2.1	9	bad	1	0.59375	0.357740586	0.217005076
2	11	bad	1	0.59375	1.744897959	1.058453331
4	15	good	1	0.59375	0.357740586	1.017857143
4.3	12	good	1	0.59375	0.357740586	0.217005076
2.5	11	bad	1	0.59375	1.744897959	1.058453331
3.5	?	good	1	0.59375	0.357740586	1.017857143
4.5	10	good	1	0.59375	1.744897959	1.058453331
6	9	good	1	0.59375	1.744897959	1.058453331
2	10	bad	1	0.59375	0.357740586	0.217005076
4.5	10	good	1	0.59375	1.744897959	1.058453331
3	12	good	1	0.59375	0.357740586	1.017857143
5	11	good	1	0.59375	0.357740586	0.217005076
2	10	bad	1	0.59375	0.357740586	0.217005076
4.5	11	good	1	0.59375	0.357740586	0.217005076
3	10	bad	1	3.166666667	1.907949791	1.157360406
2.5	10	bad	1	0.59375	0.357740586	0.217005076
4	10	bad	1	3.166666667	1.907949791	1.157360406
2	10	bad	1	0.59375	0.357740586	0.217005076
2	11	bad	1	0.59375	1.744897959	1.058453331
2	11	bad	1	0.59375	1.744897959	1.058453331
2.8	9	bad	1	3.166666667	1.907949791	1.157360406
2	10	bad	1	0.59375	0.357740586	0.217005076
4.5	12	good	1	0.59375	0.357740586	0.217005076
4	11	bad	1	3.166666667	9.306122449	5.64508443
2	12	bad	1	0.59375	1.744897959	1.058453331
2.5	12	bad	1	0.59375	1.744897959	1.058453331
2.5	11	bad	1	0.59375	1.744897959	1.058453331
4	10	bad	1	3.166666667	1.907949791	1.157360406
4.5	10	bad	1	3.166666667	1.907949791	5.428571429
4.5	11	good	1	0.59375	0.357740586	0.217005076
4.6	?	good	1	0.59375	0.357740586	0.217005076
5	11	good	1	0.59375	0.357740586	0.217005076
5.7	11	good	1	0.59375	0.357740586	0.217005076
7	11	good	1	0.59375	0.357740586	0.217005076
2	?	good	1	3.166666667	1.907949791	5.428571429
3.5	13	good	1	0.59375	0.357740586	1.017857143
4	11	good	1	0.59375	0.357740586	1.017857143
5	11	good	1	0.59375	0.357740586	0.217005076
5	12	good	1	0.59375	0.357740586	0.217005076
6	9	good	1	0.59375	1.744897959	1.058453331

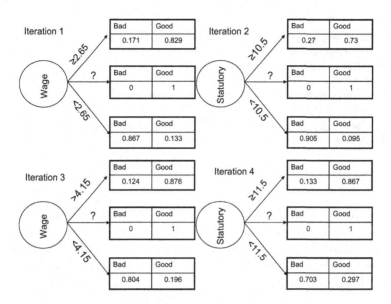

Fig. 2.11   Classifiers obtained by the first four iterations of AdaBoost.

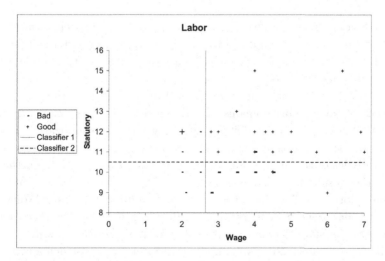

Fig. 2.12   Labor instances weights after the second iteration. Instances whose weights were increased are depicted by larger symbols. The horizontal line depicts the decision boundary that was added by the second classifier.

Input: $S = \langle (x_1, y_1), \ldots, (x_N, y_N) \rangle$, No. of Iterations T
        Loss function $G : \mathbb{R}^N \to \mathbb{R}$
Initialize: $f_0 \equiv 0$, $d_n^1 = g'(f_0(x_n), y_n)$ for all $n = 1 \ldots N$
Do for $t = 1, \ldots, T$,
(a) Train classifier on $\{S, \mathbf{d}^t\}$ and obtain hypothesis $H \in h_t : \mathbb{X} \to \mathbb{Y}$
(b) Set $\alpha_t = \mathrm{argmin}_{\alpha \in \mathbb{R}} G[f_t + \alpha h_t]$
(c) Update $f_{t+1} = f_t + \alpha_t h_t$ and $d_n^{(t+1)} = g'(f_{t+1}(x_n), y_n)$, $n = 1, \ldots, N$
Output: f_T

Fig. 2.13   Leveraging algorithm presented in Meir and Ratsch (2003).

Meir and Ratsch (2003) provide an overview of theoretical aspects of AdaBoost and its variants. They also suggest a generic scheme called leveraging and many of the boosting algorithms can be derived from this scheme. This generic scheme is presented in Figure 2.13. Different algorithms use different loss functions G.

2.6   No Free Lunch Theorem and Ensemble Learning

Empirical comparison of different ensemble approaches and their variants in a wide range of application domains has shown that each performs best in some, but not all, domains. This has been termed the *selective superiority problem* [Brodley (1995a)].

It is well known that no induction algorithm can produce the best performance in all possible domains. Each algorithm is either explicitly or implicitly biased [Mitchell (1980)] - preferring certain generalizations over others, and the algorithm is successful as long as this bias matches the characteristics of the application domain [Brazdil *et al.* (1994)]. Furthermore, other results have demonstrated the existence and correctness of the "conservation law" [Schaffer (1994)] or "no free lunch theorem" [Wolpert (1996)]: *if one inducer is better than another in some domains, then there are necessarily other domains in which this relationship is reversed.*

The "no free lunch theorem" implies that for a given problem, a certain approach can derive more information than other approaches when they are applied to the same data: "for any two learning algorithms, there are just as many situations (appropriately weighted) in which algorithm one is superior to algorithm two as vice versa, according to any of the measures of superiority." [Wolpert (2001)]

A distinction should be made between all the mathematically possible domains, which are simply a product of the representation languages used, and the domains that occur in the real world, and are therefore the ones of primary interest [Rao *et al.* (1995)]. Obviously, there are many domains in the former set that are not in the latter, and the average accuracy in the real world domains can be increased at the expense of accuracy in the domains that never occur in practice. Indeed, achieving this is the goal of inductive learning research. It is still true that some algorithms will match certain classes of naturally occurring domains better than other algorithms, and therefore achieve higher accuracy than these algorithms (this may be reversed in other real-world domains). However, this does not preclude an improved algorithm from being as accurate as the best in every domain class.

Indeed, in many application domains the generalization error of even the best methods is significantly higher than 0%, and an open and important question is whether it can be improved, and if so how. In order to answer this question, one must first determine the minimum error achievable by any classifier in the application domain (known as the optimal Bayes error). If existing classifiers do not reach this level, new approaches are needed. Although this problem has received considerable attention (see for instance [Tumer and Ghosh (1996)]), none of the methods proposed in the literature so far, is accurate for a wide range of problem domains.

The "no free lunch" concept presents a dilemma to the analyst who needs to solve a new task: which inducer should be used?

Ensemble methods overcome the no-free-lunch dilemma, by combining the outputs of many classifiers, assuming that each classifier performs well in certain domains while being sub-optimal in others. Specifically, *multi-strategy learning* [Michalski and Tecuci (1994)], attempts to combine two or more different paradigms in a single algorithm. Most research in this area has focused on combining empirical and analytical approaches (see for instance [Towell and Shavlik (1994)]. Ideally, a multi-strategy learning algorithm would always perform as well as the best of its members, there by alleviating the need to individually employ each one and simplifying the knowledge acquisition task. More ambitiously, combining different paradigms may produce synergistic effects (for instance, by constructing various types of frontiers between different class regions in the instance space), leading to levels of accuracy that no individual atomic approach can achieve.

2.7 Bias-Variance Decomposition and Ensemble Learning

It is well known that the error can be decomposed into three additive components [Kohavi and Wolpert (1996)]: the intrinsic error, the bias error and the variance error.

The intrinsic error, also known as the irreducible error, is the component that is generated due to noise. This quantity is the lower bound of any inducer, i.e. it is the expected error of the Bayes optimal classifier. The bias error of an inducer is the persistent or systematic error that the inducer is expected to make. Variance is a concept closely related to bias. The variance captures random variation in the algorithm from one training set to another. Namely ,it measures the sensitivity of the algorithm to the actual training set, or the error that is due to the training set's finite size.

The following equations provide one of the possible mathematical definitions of the various components in case of a zero-one lose:

$$t(I, S, c_j, x) = \begin{cases} 1 & \hat{P}_{I(S)}(y = c_j \,|x) > \hat{P}_{I(S)}(y = c^* \,|x) \, \forall c^* \in dom(y), \neq c_j \\ 0 & Otherwise \end{cases}$$

$$bias^2(P(y\,|x), \hat{P}_I(y\,|x)) =$$

$$\frac{1}{2} \sum_{c_j \in dom(y)} \left[P(y = c_j \,|x) - \sum_{\forall S, |S|=m} P(S\,|D) \cdot t(I, S, c_j, x) \right]^2$$

$$var(\hat{P}_I(y\,|x)) = \frac{1}{2} \left\{ 1 - \sum_{c_j \in dom(y)} \left[\sum_{\forall S, |S|=m} P(S\,|D) \cdot t(I, S, c_j, x) \right]^2 \right\}$$

$$var(P(y\,|x)) = \frac{1}{2} \left\{ 1 - \sum_{c_j \in dom(y)} [P(y = c_j \,|x)]^2 \right\}$$

Note that the probability to misclassify the instance x using inducer I and a training set of size m is:

$$\varepsilon(x) = bias^2(P(y\,|x), \hat{P}_I(y\,|x)) + var(\hat{P}_I(y\,|x)) + var(P(y\,|x))$$
$$= 1 - \sum_{c_j \in dom(y)} P(y = c_j \,|x) \cdot \sum_{\forall S, |S|=m} P(S\,|D) \cdot t(I, S, c_j, x)$$

where I is an inducer, S is a training set, c_j is a class label, x is a pattern and D is the instance domain. It is important to note that in case of a zero-one loss there are other definitions for the bias and variance components. These definitions are not necessarily consistent. In fact, there is a considerable debate in the literature about what should be the most appropriate definition. For a complete list of these definitions please refer to [Hansen (2000)].

Nevertheless, in the regression problem domain a single definition of bias and variance has been adopted by the entire community. In this case it is useful to define the bias-variance components by referring to the quadratic loss, as follows:

$$E\left(f(x) - \hat{f}_R(x)^2\right) =$$
$$\text{var}\left(f(x)\right) + \text{var}\left(\hat{f}_R(x)\right) + bias^2\left(f(x), \hat{f}_R(x)\right)$$

where $\hat{f}_R(x)$ is the prediction of the regression model and $f(x)$ is the actual value. The intrinsic variance and bias components are respectively defined as:

$$\text{var}(f(x)) = E((f(x) - E(f(x)))^2)$$
$$\text{var}(\hat{f}_R(x)) = E((\hat{f}_R(x) - E(\hat{f}_R(x)))^2)$$
$$bias^2(f(x), \hat{f}_R(x)) = E((E(\hat{f}_R(x)) - E(f(x)))^2$$

Simpler models tend to have a higher bias error and a smaller variance error than complicated models. [Bauer and Kohavi (1999)] have presented an experimental result supporting the last argument for the Naïve Bayes inducer, while [Dietterich and Kong (1995)] examined the bias-variance issue in decision trees. Figure 2.14 illustrates this argument. The figure shows that there is a tradeoff between variance and bias. When the classifier is simple it has a large bias error and a small variance error. As the complexity of the classifier increases, the variance error becomes larger and the bias error becomes smaller. The minimum generalization error is obtained somewhere in between, where the bias and variance are equal.

Empirical and theoretical evidence show that some ensemble techniques (such as bagging) act as a variance reduction mechanism, i.e., they reduce the variance component of the error. Moreover, empirical results suggest that other ensemble techniques (such as AdaBoost) reduce both the bias and the variance parts of the error. In particular, it seems that the bias error is mostly reduced in the early iterations, while the variance error decreases in later ones.

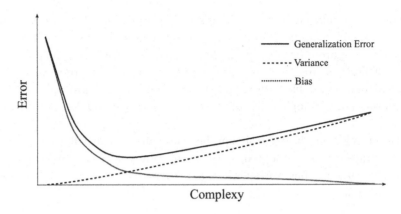

Fig. 2.14   Bias error vs. variance error in the Deterministic Case: (Hansen, 2000).

2.8   Occam's Razor and Ensemble Learning

William of Occam was a 14th-century English philosopher that presented an important principle in science which bears his name. *Occam's razor* states that the explanation of any phenomenon should make as few assumptions as possible, eliminating those that make no difference in the observable predictions of the explanatory hypothesis or theory.

According to [Domingos (1999)] there are two different interpretations of Occam's razor when it is applied to the pattern recognition domain:

- First razor – Given two classifiers with the same generalization error, the simpler one should be preferred because simplicity is desirable in itself.
- Second razor – Given two classifiers with the same training set error, the simpler one should be preferred because it is likely to have a lower generalization error.

It has been empirically observed that certain ensemble techniques often do not overfit the model, even when the ensemble contains thousands of classifiers. Furthermore, occasionally, the generalization error would continue to improve long after the training error had reached zero. Figure 2.15 illustrates this phenomenon. The figure shows the graphs of the training and test error produced by an ensemble algorithm as a function of its size. While the training error reaches zero, the test error, which approximate the generalization error, continues to reduce. This obviously contradicts the spirit of the second razor described above. Comparing the performance

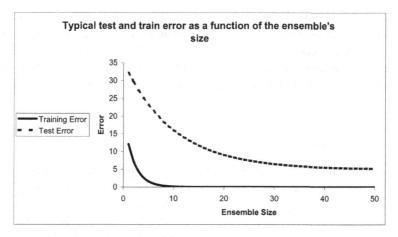

Fig. 2.15 Graphs of the training and test errors produced by an ensemble algorithm as a function of its size.

of the ensemble that contains twenty members to the one that contains thirty members, we notice that both have the same training error. Thus, according to second razor, one should prefer the simplest ensemble (i.e. twenty members). However, the graph indicates that we should, instead, prefer the largest ensemble since it obtains a lower test error. This contradiction can be settled by arguing that the first razor is largely agreed upon, while the second one, when taken literally, is false [Domingos (1999)].

Freund (2009) claims that there are two main theories for explaining the phenomena presented in Figure 2.15. The first theory relates ensemble methods such as Adaboost to logistic regression. The decrease in misclassification rate of the ensemble is seen as a by-product of the likelihood's improvement. The second theory refers to the large margins theory. Like in the theory of support vector machines (SVM), the focus of large margin theory is on the task of reducing the classification error rate on the test set.

2.9 Classifier Dependency

Ensemble methods can be differentiated according to which extent each classifier affect the other classifiers. This property indicates whether the various classifiers are dependent or independent. In a dependent framework the outcome of a certain classifier affects the creation of the next classifier. Alternatively each classifier is built independently and their results are combined in some fashion. Some researchers refer to this property as "the relationship between modules" and distinguish between three different types: successive, cooperative and supervisory [Sharkey (1996)]. Roughly speaking, "successive" refers to "dependent" while "cooperative" refers to "independent". The last type applies to those cases in which one model controls the other model.

2.9.1 Dependent Methods

In dependent approaches for learning ensembles, there is an interaction between the learning runs. Thus it is possible to take advantage of knowledge generated in previous iterations to guide the learning in the next iterations. We distinguish between two main approaches for dependent learning, as described in the following sections [Provost and Kolluri (1997)].

2.9.1.1 Model-guided Instance Selection

In this dependent approach, the classifiers that were constructed in previous iterations are used for manipulating the training set for the following iteration (see Figure 2.16). One can embed this process within the basic learning algorithm. These methods usually ignore all data instances on which their initial classifier is correct and only learn from misclassified instances.

2.9.1.2 Basic Boosting Algorithms

The most well known model-guided instance selection is boosting. Boosting (also known as arcing — Adaptive Resampling and Combining) is a general method for improving the performance of a weak learner (such as classification rules or decision trees). The method works by repeatedly running a weak learner (such as classification rules or decision trees), on various distributed training data. The classifiers produced by the weak learners are then combined into a single composite strong classifier in order to achieve a higher accuracy than the weak learner's classifiers would have had.

The AdaBoost algorithm was first introduced in [Freund and Schapire (1996)]. The main idea of this algorithm is to assign a weight in each example in the training set. In the beginning, all weights are equal, but in every round, the weights of all misclassified instances are increased while the weights of correctly classified instances are decreased. As a consequence, the weak learner is forced to focus on the difficult instances of the training

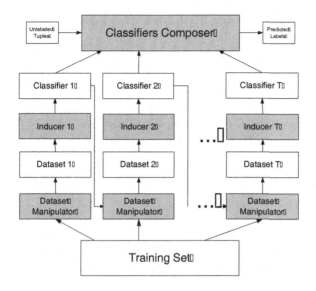

Fig. 2.16   Model guided instance selection diagram.

set. This procedure provides a series of classifiers that complement one another.

Breiman [Breiman (1998)] explores a simpler arcing algorithm called Arc-x4 which was aim to demonstrate that AdaBoost works not because of the specific form of the weighing function, but because of the adaptive resampling. In Arc-x4 the classifiers are combined by simple voting and the updated $t + 1$ iteration probabilities are defined by:

$$D_{t+1}(i) = 1 + m_{t_i}^4 \qquad (2.3)$$

where m_{t_i} is the number of misclassifications of the i-th instance by the first t classifiers.

The basic AdaBoost algorithm deals with binary classification. Freund and Schapire describe two versions of the AdaBoost algorithm (AdaBoost.M1, AdaBoost.M2), which are equivalent for binary classification and differ in their handling of multi-classes classification problems. Figure 2.17 describes the pseudo-code of AdaBoost.M1. The classification of a new instance is performed according to the following equation:

$$H(x) = \operatorname*{argmax}_{y \in dom(y)} \left(\sum_{t: M_t(x)=y} \log \frac{1}{\beta_t} \right) \qquad (2.4)$$

Require: I (a weak inducer), T (the number of iterations), S (the training set)

Ensure: $M_t, \beta_t; t = 1, \ldots, T$

1: $t \leftarrow 1$

2: $D_1(i) \leftarrow 1/m; i = 1, \ldots, m$

3: **repeat**

4:    Build Classifier M_t using I and distribution D_t

5:    $\varepsilon_t \leftarrow \displaystyle\sum_{i: M_t(x_i) \neq y_i} D_t(i)$

6:    **if** $\varepsilon_t > 0.5$ **then**

7:       $T \leftarrow t - 1$

8:       exit Loop.

9:    **end if**

10:    $\beta_t \leftarrow \frac{\varepsilon_t}{1 - \varepsilon_t}$

11:    $D_{t+1}(i) = D_t(i) \cdot \begin{cases} \beta_t & M_t(x_i) = y_i \\ 1 & Otherwise \end{cases}$

12:    Normalize D_{t+1} to be a proper distribution.

13:    $t + +$

14: **until** $t > T$

Fig. 2.17   The AdaBoost.M.1 algorithm.

where β_t is defined in Figure 2.17.

AdaBoost.M2 is a second alternative extension of AdaBoost to the multi-class case. This extension requires more elaborate communication between the boosting algorithm and the weak learning algorithm. AdaBoost.M2 uses the notion of "pseudo-loss" which measures the goodness of the weak hypothesis. The pseudocode of AdaBoost.M2 is presented in Figure 2.18. A different weight $w_{i,y}^t$ is maintained for each instance i and each label $y \in Y - \{y_i\}$. The function $q = \{1, \ldots, N\} \times Y \to [0, 1]$, called the *label weighting function*, assigns to each example i in the training set a probability distribution such that, for each i: $\sum_{y \neq y_i} q(i, y) = 1$. The inducer gets both a distribution D_t and a label weight function q_t. The inducer's target is to minimize the pseudo-loss ϵ_t for given distribution D and weighting function q.

2.9.1.3   *Advanced Boosting Algorithms*

Friedman *et al.* [Friedman *et al.* (2000)] present a generalized version of AdaBoost, which they call Real AdaBoost. The revised algorithm combines

Algorithm AdaBoost.M2

Require: I (a base inducer), T (number of iterations), S (the original training set), μ (the sample size).

1: Initialize the weight vector $D_1(i) \leftarrow 1/m$ $i = 1,\ldots,m$ and $w_y^1 = D(i)/(k-1)$ for $i = 1,\ldots,m$; $y \in Y - y_i$.

2: **for** $t = 1, 2, \ldots, T$ **do**

3:     Set $W^f = \Sigma_{y \neq y} w_{\iota,y}^t$

4:     $q_t(i, y) = \frac{w_y^f}{W^f}$ for $y \neq y_,$;

5:     Set $D_t(i) = \frac{W^f}{\Sigma_{=1}^N W^f}$.

6:     Call I, providing it with the distribution D_t and label weighting function q_t; get back a hypothesis Mt: $x \times Y \to [0,1]$.

7:     Calculate the pseudo-loss of M_t: $\epsilon, = \frac{1}{2}\sum_{=1}^N D,(i)(1 - h,(x,,y,) + \sum_{y \neq y} q,(i, y)h,(x,,y))$.

8:     Set $\beta, = \epsilon,/(1 - \epsilon,)$.

9:     Set the new weights vector to be $w_y^{r+1} = w_y^f \beta_t^{(1/2)(1+h_t(xy)-h_t(x_\ell y))}$ for $i = 1, \ldots, N$, $y \in Y - \{y,\}$.

10: **end for**

Fig. 2.18   The AdaBoost.M2 algorithm.

the class probability estimate of the classifiers by fitting an additive logistic regression model in a forward stepwise manner. The revision reduces computation cost and may lead to better performance especially in decision trees. Moreover it can provide interpretable descriptions of the aggregate decision rule.

Friedman [Friedman (2002)] developed gradient boosting which builds ensemble by sequentially fitting base learner parameters to current "pseudo"-residuals by least squares at each iteration. The pseudo-residuals are the gradient of the loss functional being minimized, with respect to the model values at each training data point evaluated at the current step. To improve accuracy performance, increase robustness and reduce computational cost, at each iteration a subsample of the training set is randomly selected (without replacement) and used to fit the base classifier.

Phama and Smeuldersb [Phama and Smeuldersb (2008)] present a strategy to improve the AdaBoost algorithm with a quadratic combination of base classifiers. The idea is to construct an intermediate learner operating on the combined linear and quadratic terms.

First a classifier is trained by randomizing the labels of the training examples. Next, the learning algorithm is called repeatedly, using a systematic update of the labels of the training examples in each round. This method is in contrast to the AdaBoost algorithm that uses reweighting of training examples. Together they form a powerful combination that makes intensive use the given base learner by both reweighting and relabeling the original training set. Compared to AdaBoost, quadratic boosting better exploits the instances space and compares favorably with AdaBoost on large datasets at the cost of training speed. Although training the ensemble takes about 10 times more than AdaBoost, the classification time for both algorithms is equivalent.

Tsao and Chang [Tsao and Chang (2007)] refer to boosting as a stochastic approximation procedure Based on this viewpoint they develop the SA-Boost (stochastic approximation) algorithm which is similar to AdaBoost except the way members' weights is calculated.

All boosting algorithms presented here assume that the weak inducers which are provided can cope with weighted instances. If this is not the case, an unweighted dataset is generated from the weighted data by a resampling technique. Namely, instances are chosen with a probability according to their weights (until the dataset becomes as large as the original training set).

AdaBoost rarely suffers from overfitting problems. Freund and Schapire (2000) note that, "one of the main properties of boosting that has made it interesting to statisticians and others is its relative (but not complete) immunity to overfitting". In addition Breiman (2000) indicates that "a crucial property of AdaBoost is that it almost never overfits the data no matter how many iterations it is run". Still in highly noisy datasets overfitting does occur.

Another important drawback of boosting is that it is difficult to understand. The resulting ensemble is considered to be less comprehensible since the user is required to capture several classifiers instead of a single classifier. Despite the above drawbacks, Breiman [Breiman (1996a)] refers to the boosting idea as the most significant development in classifier design of the Nineties.

Sun et al. [Sun et al. (2006)] pursue a strategy which penalizes the data distribution skewness in the learning process to prevent several hardest examples from spoiling decision boundaries. They use two smooth convex penalty functions, based on Kullback–Leibler divergence (KL) and l2 norm, to derive two new algorithms: AdaBoostKL and AdaBoostNorm2 . These two AdaBoost variations achieve better performance on noisy datasets.

Induction algorithms have been applied with practical success in many relatively simple and small-scale problems. However, most of these algorithms require loading the entire training set to the main memory. The need to induce from large masses of data, has caused a number of previously unknown problems, which, if ignored, may turn the task of efficient pattern recognition into mission impossible. Managing and analyzing huge datasets requires special and very expensive hardware and software, which often forces us to exploit only a small part of the stored data.

Huge databases pose several challenges:

- Computing complexity: Since most induction algorithms have a computational complexity that is greater than linear in the number of attributes or tuples, the execution time needed to process such databases might become an important issue.
- Poor classification accuracy due to difficulties in finding the correct classifier. Large databases increase the size of the search space, and this in turn increases the chance that the inducer will select an overfitted classifier that is not valid in general.
- Storage problems: In most machine learning algorithms, the entire training set should be read from the secondary storage (such as magnetic storage) into the computer's primary storage (main memory) before the induction process begins. This causes problems since the main memory's capability is much smaller than the capability of magnetic disks.

Instead of training on a very large data base, Breiman (1999) proposes taking small pieces of the data, growing a classifier on each small piece and then combining these predictors together. Because each classifier is grown on a modestly-sized training set, this method can be used on large datasets. Moreover this method provides an accuracy which is comparable to that which would have been obtained if all data could have been held in main memory. Nevertheless, the main disadvantage of this algorithm, is that in most cases it will require many iterations to truly obtain a accuracy comparable to Adaboost.

An online boosting algorithm called ivoting trains the base models using consecutive subsets of training examples of some fixed size [Breiman (1999)]. For the first base classifier, the training instances are randomly selected from the training set. To generate a training set for the kth base classifier, ivoting selects a training set in which half the instances have been correctly classified by the ensemble consisting of the previous base classifiers and half have been misclassified. ivoting is an improvement on boosting that is less vulnerable to noise and overfitting. Further, since it does not require weighting the base classifiers, ivoting can be used in a parallel fashion, as demonstrated in [Chawla *et al.* (2004)].

Merler *et al.* [Merler *et al.* (2007)] developed the P-AdaBoost algorithm which is a distributed version of AdaBoost. Instead of updating the "weights" associated with instance in a sequential manner, P-AdaBoost works in two phases. In the first phase, the AdaBoost algorithm runs in its sequential, standard fashion for a limited number of steps. In the second phase the classifiers are trained in parallel using weights that are estimated from the first phase. P-AdaBoost yields approximations to the standard AdaBoost models that can be easily and efficiently distributed over a network of computing nodes.

Zhang and Zhang [Zhang and Zhang (2008)] have recently proposed a new boosting-by-resampling version of Adaboost. In the local Boosting algorithm, a local error is calculated for each training instance which is then used to update the probability that this instance is chosen for the training set of the next iteration. After each iteration in AdaBoost, a global error measure is calculated that refers to all instances. Consequently noisy instances might affect the global error measure, even if most of the instances can be classified correctly. Local boosting aims to solve this problem by inspecting each iteration locally, per instance. A local error measure is calculated for each instance of each iteration, and the instance receives a score, which will be used to measure its significance in classifying new instances. Each instance in the training set also maintains a local weight, which controls its chances of being picked in the next iteration. Instead of automatically increasing the weight of a misclassified instance (like in AdaBoost), we first compare the misclassified instance with similar instances in the training set. If these similar instances are classified correctly, the misclassified instance is likely to be a noisy one that cannot contribute to the learning procedure and thus its weight is decreased. If the instance's neighbors are also misclassified, the instance's weight is increased. As in AdaBoost, if an instance is classified correctly, its weight is decreased. Classifying a new instance is based on its similarity with each training instance.

The advantages of local boosting compared to other ensemble methods are:

(1) The algorithm tackles the problem of noisy instances. It has been

empirically shown that the local boosting algorithm is more robust to noise than Adaboost.

(2) In respect to accuracy, LocalBoost generally outperforms Adaboost. Moreover, LocalBoost outperforms Bagging and Random Forest when the noise level is small

The disadvantages of local boosting compared to other ensemble methods are:

(1) When the amount of noise is large, LocalBoost sometimes performs worse than Bagging and Random Forest.
(2) Saving the data for each instance increases storage complexity; this might confine the use of this algorithm to limited training sets.

AdaBoost.M1 algorithm guaranties an exponential decrease of an upper bound of the training error rate as long as the error rates of the base classifiers are less than 50%. For multiclass classification tasks, this condition can be too restrictive for weak classifiers like decision stumps. In order to make AdaBoost.M1 suitable to weak classifiers, BoostMA algorithm modifies it by using a different function to weight the classifiers [Freund (1995)]. Specifically, the modified function becomes positive if the error rate is less than the error rate of default classification. As opposed to AdaBoost.M2, where the weights are increased if the error rate exceeds 50%, in BoostMA the weights are increased for instances for which the classifier performed worse than the default classification (i.e. classification of each instance as the most frequent class). Moreover in BoostMA the base classifier minimizes the confidence-rated error instead of the pseudo-loss / error-rate (as in AdaBoost.M2 or Adaboost.M1) which makes it easier to use with already existing base classifiers.

AdaBoost-r is a variant of AdaBoost which considers not only the last weak classifier, but a classifier formed by the last r selected weak classifiers (r is a parameter of the method). If the weak classifiers are decision stumps, the combination of r weak classifiers is a decision tree. A primary drawback of AdaBoost-r is that it will only be useful if the classification method does not generate strong classifiers.

Figure 2.19 presents the pseudocode of AdaBoost-r. In line 1 we initialize a distribution on the instances so that the sum of all weights is 1 and all instances obtain the same weight. In line 3 we perform a number of iterations according to the parameter T. In lines 4-8 we define the training set S' on which the base classifier will be trained. We check whether the resampling or reweighting version of the algorithm is required. If resampling was chosen, we perform a resampling of the training set according to the distribution D_t. The resampled set S' is the same size as S. However, the instances it contains were drawn from S with repetition with the probabilities according to D_t. Otherwise (if reweighting was chosen) we simply set S' to be the entire original dataset S. In line 9 we train a base classifier M_t from the base inducer I on S' while using as instance weights the values of the distribution D_t. In line 10 lies the significant change of the algorithm compared to the normal AdaBoost. Each of the instances of the original dataset S is classified with the base classifier M_t. This most recent classification is saved in the sequence of the last R classifications. Since we are dealing with a binary class problem, the class can be represented by a single bit (0 or 1). Therefore the sequence can be stored as a binary sequence with the most recent classification being appended as the least significant bit. This is how past base classifiers are combined (through the classification sequence). The sequence can be treated as a binary number representing a leaf in the combined classifier M_t^r to which the instance belongs. Each leaf has two buckets (one for each class). When an instance is assigned to a certain leaf, its weight is added to the bucket representing the instance's real class. Afterwards, the final class of each leaf of M_t^r is decided by the heaviest bucket. The combined classifier does not need to be explicitly saved since it is represented by the final classes of the leaves and the base classifiers $M_t,\ M_{t-1},\ldots M_{max(t-r,1)}$. In line 11 the error rate ε_t of M_t^r on the original dataset S is computed by summing the weight of all the instances the combined classifier has misclassified and then dividing the sum by the total weight of all the instances in S. In line 12 we check whether the error rate is over 0.5 which would indicate the newly combined classifier is even worse than random or the error rate is 0 which indicates overfitting. In case resampling was used and an error rate of 0 was obtained, it could indicate an unfortunate resampling and so it is recommended to return to the resampling section (line 8) and retry (up to a certain number of failed attempts, e.g. 10). Line 15 is executed in case the error rate was under 0.5 and therefore we define α_t to be $\frac{1-\varepsilon_t}{\varepsilon_t}$. In lines 16-20 we iterate over all of the instances in S and update their weight for the next iteration (D_{t+1}). If the combined classifier has misclassified the instance, its weight is multiplied by α_t. In line 21, after the weights have been updated , they are renormalized so that D_{t+1} will be a distribution (i.e. all weights will sum to 1). This concludes the iteration and everything is ready for the next iteration.

For classifying an instance, we traverse each of the combined classifiers, classify the instance with it and receive either -1 or 1. The class is then multiplied by $\log(\alpha_t)$, which is the weight assigned to the classifier trained at iteration t, and added to a global sum. If the sum is positive, the class "1" is returned; if it is negative, "-1" is returned; and if it is 0, the returned class is random. This can also be viewed as summing the weights of the classifiers per class and returning the class with the maximal sum. Since we do not explicitly save the combined classifier M_t^r, we obtain its classification by classifying the instance with the relevant base classifiers and using the binary classification sequence which is given by $(M_t(x), M_{t-1}(x), \ldots M_{\max(t-r,1)}(x))$ as a leaf index into the combined classifier and using the final class of the leaf as the classification result of M_t^r.

AdaBoost.M1 is known to have problems when the base classifiers are weak, i.e. the predictive performance of each base classifier is not much higher than that of a random guessing.

AdaBoost.M1W is a revised version of AdaBoost.M1 that aims to improve its accuracy in such cases (Eibl and Pfeiffer, 2002). The required revision results in a change of only one line in the pseudo-code of AdaBoost.M1. Specifically the new weight of the base classifier is defined as:

$$\alpha_t = \ln\left(\frac{(|dom(y)| - 1)(1 - \varepsilon_t)}{\varepsilon_t}\right) \tag{2.5}$$

where ε_t is the error estimation which is defined in the original AdaBoost.M1 and $|dom(y)|$ represents the number of classes. Note the above equation generalizes AdaBoost.M1 by setting $|dom(y)| = 2$.

2.9.1.4 *Incremental Batch Learning*

In this method the classification produced in one iteration is given as "prior knowledge" to the learning algorithm in the following iteration. The learning algorithm uses the current training set together with the classification of the former classifier for building the next classifier. The classifier constructed at the last iteration is chosen as the final classifier.

2.9.2 *Independent Methods*

In this methodology the original dataset is partitioned into several subsets from which multiple classifiers are induced. Figure 2.20 illustrates the independent ensemble methodology. The subsets created from the origi-

AdaBoost-r Algorithm

Require: I (a base inducer), T (number of iterations), S (the original training set), ρ (whether to perform resampling or reweighting), r (reuse level).

1: Initiate: $D_1(X_i) = \frac{1}{m}$ for all i's.

2: $T = 1$

3: **repeat**

4:     **if** ρ **then**

5:         $S' = resample\,(S, D_t)$

6:     **else**

7:         $S' = S$

8:     **end if**

9:     Train base classifier M_t with I on S' with instance weights according to D_t

10:     Combine classifiers $M_t,\ M_{t-1}, \dots M_{max(t-r,1)}$ to create M_t^r.

11:     Calculate the error rate ε_t of the combined classifier M_t^r on S

12:     **if** $\varepsilon_t \geq 0.5$ or $\varepsilon_t = 0$ **then**

13:         End;

14:     **end if**

15:     $\alpha_t = \frac{1-\varepsilon_t}{\varepsilon_t}$

16:     **for** i=1 to m **do**

17:         **if** $M_t^r(\,X_i) \neq Y_i$ **then**

18:             $D_{t+1}(X_i) = D_t(X_i)\,\alpha_t$

19:         **end if**

20:     **end for**

21:     Renormalize D_{t+1} so it will be a distribution

22:     $t \leftarrow t + 1$

23: **until** $t > T$

Fig. 2.19   AdaBoost-r algorithm.

nal training set may be disjointed (mutually exclusive) or overlapping. A combination procedure is then applied in order to produce a single classification for a given instance. Since the method for combining the results of induced classifiers is usually independent of the induction algorithms, it can be used with different inducers at each subset. Moreover this methodology can be easily parallelized. These independent methods aim either at improving the predictive power of classifiers or decreasing the total execution time. The following sections describe several algorithms that implement

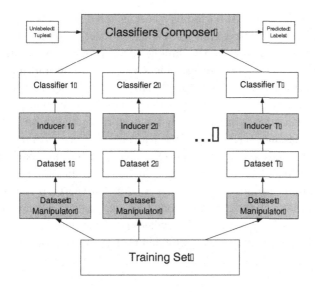

Fig. 2.20 Independent methods.

this methodology.

2.9.2.1 *Bagging*

The most well-known independent method is bagging (bootstrap aggregating) [Breiman (1996a)]. The method aims to increase accuracy by creating an improved composite classifier, by amalgamating the various outputs of learned classifiers into a single prediction. Each classifier is trained on a sample of instances taken with a replacement from the training set. Each sample size is equal to the size of the original training set. Note that since sampling with replacement is used, some of the original instances may appear more than once in the each training set and some may not be included at all.

So while the training sets may be different from each other, they are certainly not independent from a statistical point of view. To classify a new instance, each classifier returns the class prediction for the unknown instance. The composite bagged classifier, returns the class that has been predicted most often (voting method). The result is that bagging produces a combined model that often performs better than the single model built from the original single data. Breiman [Breiman (1996a)] notes that this is true especially for unstable inducers because bagging can eliminate their instability. In this context, an inducer is considered unstable if perturbing the learning set can cause significant changes in the constructed classifier.

Bagging, like boosting, is a technique for improving the accuracy of a classifier by producing different classifiers and combining multiple models. They both use a kind of voting for classification in order to combine the outputs of the different classifiers of the same type. In boosting, unlike bagging, each classifier is influenced by the performance of those built before with the new classifier trying to pay more attention to errors that were made in the previous ones and to their performances. In bagging, each instance is chosen with equal probability, while in boosting, instances are chosen with a probability proportional to their weight. Furthermore, according to Quinlan [Quinlan (1996)], as mentioned above, bagging requires that the learning system should not be stable, where boosting does not preclude the use of unstable learning systems, provided that their error rate can be kept below 0.5.

2.9.2.2 *Wagging*

Wagging is a variant of bagging [Bauer and Kohavi (1999)] in which each classifier is trained on the entire training set, but each instance is stochastically assigned a weight. Figure 2.21 presents the pseudo-code of the wagging algorithm.

In fact bagging can be considered to be wagging with allocation of weights from the Poisson distribution (each instance is represented in the sample a discrete number of times). Alternatively it is possible to allocate the weights from the exponential distribution, because the exponential distribution is the continuous valued counterpart to the Poisson distribution [Webb (2000)].

Require: I (an inducer), T (the number of iterations), S (the training set), d (weighting distribution).
Ensure: $M_t; t = 1, \ldots, T$
 1: $t \leftarrow 1$
 2: **repeat**
 3:     $S_t \leftarrow S$ with random weights drawn from d.
 4:     Build classifier M_t using I on S_t
 5:     $t + +$
 6: **until** $t > T$

Fig. 2.21   The Wagging algorithm.

2.9.2.3 *Random Forest and Random Subspace Projection*

A Random Forest ensemble [Breiman (2001)] uses a large number of individual, unpruned decision trees. The individual trees are constructed using a simple algorithm presented in Figure 2.22. The *IDT* in Figure 2.22 represents any top-down decision tree induction algorithm [Rokach and Maimon (2001)] with the following modification: the decision tree is not pruned and at each node, rather than choosing the best split among all attributes, the inducer randomly samples N of the attributes and choose the best split from among those variables. The classification of an unlabeled instance is performed using majority vote.

Originally, the random forests algorithm applies only to building decision trees, and is not applicable to all types of classifiers, because it involves picking a different subset of the attributes in each node of the tree. Nevertheless, the main step of the random forest algorithm can be easily replaced with the broader "random subspace method" [Ho (1998)], which can be applied to many other inducers, such as nearest neighbor classifiers [Ho (1998)] or linear discriminators [Skurichina and Duin (2002)].

Require: *IDT* (a decision tree inducer), T (the number of iterations), S (the training set), μ (the subsample size). N (number of attributes used in each node)

Ensure: $M_t; t = 1, \ldots, T$

1: $t \leftarrow 1$
2: **repeat**
3:    $S_t \leftarrow$ Sample μ instances from S with replacement.
4:    Build classifier M_t using $IDT(N)$ on S_t
5:    $t + +$
6: **until** $t > T$

Fig. 2.22 The Random Forest algorithm.

One important advantage of the random forest method is its ability to handle a very large number of input attributes [Skurichina and Duin (2002)]. Another important feature of the random forest is that it is fast.

There are other ways to obtain random forests. For example, instead of using all the instances to determine the best split point for each feature, a sub-sample of the instances is used [Kamath and Cantu-Paz (2001)]. This sub-sample varies with the feature. The feature and split value that optimize the splitting criterion are chosen as the decision at that node. Since the split made at a node is likely to vary with the sample selected, this technique results in different trees which can be combined in ensembles.

Another method for randomization of the decision tree through histograms was proposed in [Kamath *et al.* (2002)]. The use of histograms has long been suggested as a way of making the features discrete, while reducing the time to handle very large datasets. Typically, a histogram is created for each feature, and the bin boundaries used as potential split points. The randomization in this process is expressed by selecting the split point randomly in an interval around the best bin boundary.

Using an extensive simulation study, Archer and Kimes [Archer and Kimes (2008)] examine the effectiveness of Random Forest variable importance measures in identifying the true predictor among a large number of candidate predictors. They concluded that the Random Forest technique is useful in domains which require both an accurate classifier and insight regarding the discriminative ability of individual attribute (like in microarray studies).

2.9.2.4 *Non-Linear Boosting Projection (NLBP)*

Non-Linear Boosting Projection (NLBP) combines boosting and subspace methods as follows [Garcia-Pddrajas *et al.* (2007)]:

(1) All the classifiers receive as input all the training instances for learning. Since all are equally weighted, we are not placing more emphasis on misclassified instances as in boosting.
(2) Each classifier uses a different nonlinear projection of the original data onto a space of the same dimension.
(3) The nonlinear projection is based on the projection made by the hidden neuron layer of a multilayer perceptron neural network.
(4) Following the basic principles of boosting, each nonlinear projection is constructed in order to make it easier to classify difficult instances.

Figure 2.23 presents the pseudocode of the NLBP algorithm. In lines 1-2 we convert the original data set (S) to a standard data set for multilayer perceptron learning (S'). It includes a transformation from nominal attributes to binary attributes (one binary attribute for each nominal value) and a normalization of all numeric (and binary) attributes to the $[-1, 1]$ range. In line 3 we construct the first classifier M_0, using the base inducer (I) and the converted data set S'. In lines 4-14 we construct the remaining $T - 1$ classifiers.

Each iteration includes the following steps. In lines 5-9 we try to classify each instance in S' with the previous classifier. S'' stores all the instances that we didn't classify correctly. In lines 10-11 we train a multilayer perceptron neural network using S''. The hidden layer of the network consists of the same number of hidden neurons as input attributes in S'' . After training the network we retrieve the input weights of the hidden neurons, and store them in a data structure with the iteration index (ProjectionArray). In line 12 we project all the instances of S' with the projection P to get S'''. In line 13 we construct the current classifier using the converted and projected data set (S''') and the base inducer (I). In lines 15-16 we convert the instance (nominal to binary and normalization of attributes) as we did to the training data set. The final classification is a simple majority vote.

The advantages of NLBP compared to other ensemble methods are:

(1) Experiments comparing NLBP to popular ensemble methods (Bagging, AdaBoost, LogitBoost, Arc-x4) using different base classifiers (C4.5, ANN, SVM) show very good results for NLBP.
(2) Analysis of bagging, boosting, and NLBP by the authors of the paper suggests that: *"Bagging provides diversity, but to a lesser degree than boosting. On the other hand, boosting's improvement of diversity has the side-effect of deteriorating accuracy. NLBP behavior is midway between these two methods. It is able to improve diversity, but to a lesser degree than boosting, without damaging accuracy as much as boosting. This behavior suggests that the performance of NLBP in noisy problems can be better than the performance of boosting methods.*

The drawbacks of NLBP compared to other ensembles methods are:

(1) The necessity of training an additional neuronal network for each iteration increases the computational complexity of the algorithm compared to other approaches.
(2) Using a neuronal network for projection may increase the dimensionality of the problem. Every nominal attribute is transformed to a set of binary ones.
(3) The classifiers that are constructed by this approach use a different set of attributes than the original ones. Since these new attribute lose the meaning of the original ones, it is difficult to understand the meaning of the constructed models in terms of the original domain.

NLBP - Building the ensemble

Require: I (a base inducer), T (number of iterations), S (the original training set).
1: $S* = $ A nominal to binary transfomation of S
2: $S' = $ A nomalization of $S*$
3: $M_1 = I(S')$
4: **for** $t = 2$ to T **do**
5:     $S'' = \emptyset$
6:     **for each** $x_j \in S'$ **do**
7:         **if** $M_{t-1}(x_j) \neq y_j$ **then**
8:             $S'' = S'' \bigcup \{Xj\}$
9:         **end if**
10:     **end for**
11:     Train network H using S'' and get the projection $P(X)$ implemented by hidden layer of H
12:     $ProjectionArray[t] = P$
13:     $S''' = P(S')$
14:     $M_t = I(S''')$
15: **end for**

Fig. 2.23   NLBP - Building the ensemble.

2.9.2.5 *Cross-validated Committees*

This procedure creates k classifiers by partitioning the training set into k-equal-sized sets and training, in turn, on all but the i-th set. This method, first used by Gams [Gams (1989)], employed 10-fold partitioning. Parmanto et al. [Parmanto et al. (1996)] have also used this idea for creating an ensemble of neural networks. Domingos [Domingos (1996)] used cross-validated committees to speed up his own rule induction algorithm RISE, whose complexity is $O(n^2)$, making it unsuitable for processing large databases. In this case, partitioning is applied by predetermining a maximum number of examples to which the algorithm can be applied at once. The full training set is randomly divided into approximately equal-sized partitions. RISE is then run on each partition separately. Each set of rules grown from the examples in partition p is tested on the examples in partition $p + 1$, in order to reduce overfitting and to improve accuracy.

2.9.2.6 *Robust Boosting*

A robust classifier is one whose predictive performance is not sensitive to changes in the training data. Freund (1995) proposes a simple robust version of the boosting algorithm which is called the boost-by-majority (BBM) algorithm. In BBM, the number of iterations is set in advance based on two parameters that are specified by the user: the desired accuracy and an error bound such that the induction algorithm is guaranteed to always generate a classifier whose error is smaller than that value. The main idea of BBM is to give a small weight to instances with large negative margins. Intuitively, it ignores instances which are unlikely to be classified correctly when the boosting procedure terminates. Contrary to AdaBoost, BBM assigns the weights to the base classifier regardless of their accuracies. Thus, the drawback of BBM is that it is not adaptive. On the other hand, the AdaBoost algorithm is sensitive to noise. Specifically, the accuracy of Adaboost decreases rapidly when random noise is added to the training set.

In order to avoid these drawbacks, Freund (2001) proposes BrownBoost which is adaptive version of BBM in which classifiers with a small misclassification rate are assigned a larger weight than base classifiers with large misclassification rate. BrownBoost has a time variable denoted as t that increases with each iteration. Figure 2.24 specifies the pseudocode of BrownBoost.

Most recently, Freund (2009) proposes an updated version of Brown-Boost called RobustBoost. The main difference is that instead of minimizing the training error its goal is to minimize the number of examples whose normalized margins is smaller than some value $\theta > 0$:

$$\frac{1}{N} \sum_{i=1}^{N} 1 \left[\bar{m} \left(x_i, y_i \right) \leq \theta \right] \tag{2.6}$$

where $\bar{m}(x, y)$ is the normalized margin defined as:

$$\bar{m}(x, y) = \frac{y \cdot \text{sign} \left(\sum_i \alpha_i C_i \left(x \right) \right)}{\sum_i |\alpha_i|} \tag{2.7}$$

Friedman *et al.* (2000) show that Adaboost is approximating a stepwise additive logistic regression model by optimizing an exponential criterion. Based on this observation, Friedman *et al.* (2000) propose a variant of Adaboost, called Logitboost, which fits additive models directly. Since it

BrownBoost Training

Require: I (a base inducer), S (the original training set), T (number of iterations).

1: Set initial weights as $w_i = 1/N$
2: Set $F(x) = 0$
3: **for** $t = 1$ to T **do**
4:    Fit the function f_t by a weighted least-squares regression of Z_i to x_i with weights w_i.
5:    Set $F(x) = F(x) + f_t(x)$
6:    Set $w_i \leftarrow w_i e^{-y_i f_t(x_i)}$
7: **end for**

Fig. 2.24   The BrownBoost algorithm.

uses Newton-like steps to optimize the binomial log-likelihood criterion, it is significantly better than Adaboost at tolerating noise. Despite such claims, Mease and Wyner (2008) indicate that when the Bayes error is not zero, LogitBoost often overfits while AdaBoost does not. In fact Mease and Wyner (2008) encourage the readers to try the simulation models provided on the web page http://www.davemease.com/contraryevidence. Other closely related algorithms are the log-loss Boost [Collins *et al.* (2002)] and MAdaboost [Domingo and Watanabe (2000)]. The pseudocode of LogitBoost is presented in Figure 9-6.

Zhang's boosting algorithm (Zhang *et al.*, 2009) is an Adaboost variant with the following differences: (a) Instead of using the entire dataset, a subsample of the original training data in each iteration trains the weak classifier. (b) The sampling distribution is set differently to overcome Adaboost sensitivity to noise. Specifically, a parameter introduced into the reweighted scheme proposed in Adaboost updates the probabilities assigned to training examples. The results of these changes are better prediction accuracy, faster execution time and robustness to classification noise. Figure 2.25 presents the pseudocode of this variant. The sampling introduces randomness into the procedure. Using the f parameter, one can control the amount of data available to train the weak classifier. The parameter β is used to alleviate AdaBoost's problem in which more and more weight is assigned to noisy examples in later iterations. Thus, the weight increment of inaccurately predicted examples is smaller than that in Adaboost.

Zhang's Boosting Algorithm

Require: I (a base inducer), T (number of iterations), S (the original training set), sample fraction f and positive parameter β

1: Initialize: set the probability distribution over S as $D_1(i) = 1/N (i = 1, 2, \cdots N)$

2: **for** $t = 1, \cdots T$ **do**

3: According to the distribution D_t, draw $\overline{N} = \lfloor f \cdot N \rfloor (f \leq 1)$ examples from S with replacement to compose a new training set $S_t = \{(\mathbf{x}_i^{(t)}, y_i^{(t)})\}_{i=1}^{\overline{N}}$ in which $\lfloor A \rfloor$ stands for the largest integer small than A.

4: Apply I to S_t to train a weak classifier $h_t : X \rightarrow \{-1, +1\}$ and compute the error of h_t as $\epsilon_t = \displaystyle\sum_{i.h_t(\mathbf{x}_l) \neq y_l}^{N} D_t(i)$.

5: **if** $\epsilon i_t > 0.5$ **then**

6: set $T = t - 1$ and abort loop.

7: **end if**

8: choose $\alpha_t = \dfrac{1}{2}\ln(\dfrac{1 - \epsilon_t}{\epsilon_t})$.

9: Update the probability distribution over S as $D_{t+1}(i) = \dfrac{D_t(i)}{Z_t} \times$
$$\begin{cases} e^{-\alpha_t/\beta}, \text{if} h_t(\mathbf{x}_i) = y_i \\ e^{\alpha_t/\beta}, \text{if} h_t(\mathbf{x}_i) \neq y_i \end{cases} = \frac{D_t(i)\exp((-\frac{\alpha_t}{\beta})y_i h_t(\mathbf{x}_i))}{Z_t} \text{ where } Z_t \text{ is a}$$
normalization factor (it should be chosen so that D_{t+1} is a distribution over S).

10: **end for**

Fig. 2.25 Zhang's boosting algorithm.

2.10 Ensemble Methods for Advanced Classification Tasks

2.10.1 *Cost-Sensitive Classification*

AdaBoost does not differentiate between the various classes. Thus, a misclassification in the majority class is treated equally as a misclassification of the minority class. However, in certain scenarios it is more desirable to augment the weight of misclassification errors of the minority class. For example, in direct marketing scenarios, firms are interested in estimating customer interest in their offer. However, positive response rates are usually low. For example, a mail marketing response rate of 2

$$D^{t+1}(i) = \frac{D^t(i)\sqrt{\frac{\sum_{[i,M_t(x_i)\neq y_i]}\delta\cdot W_i}{\sum_{[i,M_t(x_i)=y_i]}\delta\cdot W_i}}}{Z_t} \tag{2.8}$$

For unsuccessful classification, the distribution update is revised to:

$$D^{t+1}(i) = \frac{D^t(i)\Big/\sqrt{\frac{\sum_{[i,M_t(x_i)\neq y_i]}\delta\cdot W_i}{\sum_{[i,M_t(x_i)=y_i]}\delta\cdot W_i}}}{Z_t} \tag{2.9}$$

where Z_t is a normalization factor.

Fan *et al.* (1999) presented AdaCost. The purpose of AdaCost is to improve AdaBoosts fixed and variable misclassification costs. It introduces a cost-adjustment function which is integrated into the weight updating rule. In addition to assigning high initial weights to costly instances, the weight updating rule takes cost into account and increases the weights of costly misclassification. Figure 2.26 presents the pseudocode of AdaCost. where $\beta(i) = \beta(\text{sign}(y_i h_t(x_i)), c_i)$ is a cost-adjustment function. Z_t is a normalization factor chosen so that D_{t+1} will be a distribution. The final classification is: $H(x) = \text{sign}(f(x))$ where $f(x) = (\sum_{t=1}^{T} \alpha_t h_t(x))$

AdaCost

Require: I (a base inducer), T (number of iterations), $S = \{(x_1,\ c_1,\ y_1),\ \ldots,\ (x_m,\ c_m,\ y_m)\}$. $x_i \in \mathcal{X}, c_i \in \mathbb{R}^+,\ y_i \in \{-1, +1\}$

1: Initialize $D_1(i)(T$: such as $D_1(i) = c_i / \sum_j^m c_j)$.
2: **repeat**
3:    Train weak inducer using distribution D_t.
4:    Compute weak classifier h_t: $\mathcal{X} \to \mathbb{R}$.
5:    Choose $\alpha_t \in \mathbb{R}$ and $\beta(i) \in \mathbb{R}^+$.
6:    Update $D_{t+1}(i) = \frac{D_t(i)\exp(-\alpha_t y_i h_t(x_i)\beta(i))}{Z_t}$
7:    $t \leftarrow t + 1$
8: **until** $t > T$

Fig. 2.26   AdaCost algorithm.

2.10.2 *Ensemble for Learning Concept Drift*

Concept drift is an online learning task in which concepts change or drift over time. More specifically, concept drift occurs when the class distribution changes over time.

Concept drift exists in many applications that involve models of human behavior, such as recommender systems. Kolter and Maloof (2007) suggested an algorithm that tries to solve this problem by presenting an ensemble method for concept drift that dynamically creates and removes weighted experts according to a change in their performance. The suggested algorithm known as dynamic weighted majority (DWM) is an extension of the weighted majority algorithm (MWA) but it adds and removes base learners in response to global and local performance. As a result, DWM is better able to respond in non-stationary environments than other algorithms, especially those that rely on an ensemble of unweighted learners (such as SEA). The main disadvantage of DWM is its poor performance in terms of running time, compared to the AdaBoost algorithm.

2.10.3 *Reject Driven Classification*

Reject driven classification[Frelicot and Mascarilla (2001)] is a method of classification that allows a tradeoff between misclassification and ambiguity (assigning more than one class to an instance). Specifically, the algorithm introduces a method for combining reject driven classifiers using belief theory methods. The algorithm adjusts the results of the reject driven classifiers by using the Dempster-Shafer theory. For each classifier, a basic probability assignment (BPA) is calculated to classify unseen instances.

The main strength of this algorithm is its ability to control the tradeoff between ambiguity and rejection. We can decide (with the proper threshold) if we prefer to classify an unseen instance to a single class and might be wrong or give an ambiguity classification. A major drawback of the algorithm is its inability to handle datasets with many classes since the BPA calculation needs to calculate the probability for any pair of classes.

Chapter 3

Ensemble Classification

Ensemble classification refers to the process of using ensemble's classifiers in order to provide a single and unified classification to an unseen instance.

There are two major ways for classifying new instances. In the first approach the classification are *fused* in some fashion during the classification phase. In the second approach the classification of one classifier is *selected* according to some criterion.

3.1 Fusions Methods

Fusing methods aim at providing the classification by combining the outputs of several classifiers. We assume that the output of each classifier i is a k-long vector $p_{i,1}, \cdots, p_{i,k}$. The value $p_{i,j}$ represents the support that instance x belongs to class j according to the classifier i. For the sake of simplicity, it is also assumed that $\sum_{j=1}^{k} p_{i,j} = 1$. If we are dealing with a crisp classifier i, which explicitly assigns the instance x to a certain class l, then it can still be converted to k-long vector $p_{i,1}, \cdots, p_{i,k}$ such that $p_{i,l} = 1$ and $p_{i,j} = 0 \forall j \neq l$.

Fusions methods can be furthered partitioned into weighting methods and meta-learning methods. The following sections specify each of these techniques.

3.1.1 *Weighting Methods*

The base members classification are combined using weights that are assigned to each member. The member's weight indicates its effect on the final classification. The assigned weight can be fixed or dynamically determined for the specific instance to be classified.

The weighting methods are best suited for problems where the individual classifiers perform the same task and have comparable success or when we would like to avoid problems associated with added learning (such as overfitting or long training time).

3.1.2 *Majority Voting*

In this combining scheme, a classification of an unlabeled instance is performed according to the class that obtains the highest number of votes (the most frequent vote). This method is also known as the plurality vote (PV) or the basic ensemble method (BEM). This approach has frequently been used as a combining method for comparing newly proposed methods.

For example we are given an ensemble of ten classifiers which are attempting to classify a certain instance x to one of the classes: A, B or C. Table 3.1 presents the classification vector and the selected label (vote) that each classifier provides to the instance x. Based on these classifications we create the voting table presented in Table 3.1 which indicates that the mojarity vote is class B.

Table 3.1 Illustration of Majority Voting: Classifiers Output.

Classifier	A score	B score	C score	Selected Label
1	0.2	0.7	0.1	B
2	0.1	0.1	0.8	C
3	0.2	0.3	0.5	C
4	0.1	0.8	0.1	B
5	0.2	0.6	0.2	B
6	0.6	0.3	0.1	A
7	0.25	0.65	0.1	B
8	0.2	0.7	0.1	B
9	0.2	0.2	0.8	C
10	0.4	0.3	0.3	A

Table 3.2 Illustration of Majority Voting.

	Class A	Class B	Class C
Votes	2	5	3

Mathematically majority voting can be written as:

$$class(x) = \underset{c_i \in dom(y)}{\arg\max} \left(\sum_k g\left(y_k(x), c_i\right) \right) \tag{3.1}$$

where $y_k(x)$ is the classification of the k'th classifier and $g(y, c)$ is an indicator function defined as:

$$g(y, c) = \begin{cases} 1 & y = c \\ 0 & y \neq c \end{cases} \tag{3.2}$$

Note that in case of a probabilistic classifier, the crisp classification $y_k(x)$ is usually obtained as follows:

$$y_k(x) = \underset{c_i \in dom(y)}{\arg\max} \hat{P}_{M_k}(y = c_i \,|x) \tag{3.3}$$

where M_k denotes classifier k and $\hat{P}_{M_k}(y = c \,|x)$ denotes the probability of y obtaining the value c given an instance x.

3.1.3 *Performance Weighting*

The weight of each classifier can be set proportional to its accuracy performance on a validation set [Opitz and Shavlik (1996)]:

$$w_i = \frac{(\alpha_i)}{\sum\limits_{j=1}^{T}(\alpha_j))} \tag{3.4}$$

whore α_i io a pcrformancc cvaluation of classifier i on a validation set (for example the accuracy). Once the weights for each classifier have been computed, we select the class which receive the highest score:

$$class(x) = \underset{c_i \in dom(y)}{\arg\max} \left(\sum_k \alpha_i g\left(y_k(x), c_i\right) \right) \tag{3.5}$$

Since the weights are normalized and are summed up to 1, it possible to interpret the sum in last equation as the probability that x_i is classified into c_j.

Moreno-Seco *et al.* (2006) examined several variations of performance weighting methods:

Re-scaled weighted vote The idea is to weight values proportionally to some given ratio N/M as following:

$$\alpha_k = \max\left\{1 - \frac{M \cdot e_k}{N \cdot (M-1)}, 0\right\}$$

where e_i is the number of misclassifications made by classifier i.

Best-worst weighted vote The idea is that the best and the worst classifiers obtain the weight of 1 and 0 respectively. The rest of classifiers are rated linearly between these extremes:

$$\alpha_i = 1 - \frac{e_i - \min_i (e_i)}{\max_i (e_i) - \min_i (e_i)}$$

Quadratic best-worst weighted vote In order to give additional weight to the classifications provided by the most accurate classifiers, the values obtained by the best-worst weighted vote approach are squared:

$$\alpha_i = \left(\frac{\max_i (e_i) - e_i}{\max_i (e_i) - \min_i (e_i)}\right)^2$$

3.1.4 *Distribution Summation*

The idea of the distribution summation combining method is to sum up the conditional probability vector obtained from each classifier [Clark and Boswell (1991)]. The selected class is chosen according to the highest value in the total vector. Mathematically, it can be written as:

$$Class(x) = \underset{c_i \in dom(y)}{\operatorname{argmax}} \sum_k \hat{P}_{M_k}(y = c_i \,|x) \tag{3.6}$$

3.1.5 *Bayesian Combination*

In the Bayesian combination method the weight associated with each classifier is the posterior probability of the classifier given the training set [Buntine (1990)].

$$Class(x) = \underset{c_i \in dom(y)}{\operatorname{argmax}} \sum_k P(M_k \,|S) \cdot \hat{P}_{M_k}(y = c_i \,|x) \tag{3.7}$$

where $P(M_k \,|S)$ denotes the probability that the classifier M_k is correct given the training set S. The estimation of $P(M_k \,|S)$ depends on the classifier's representation. To estimate this value for decision trees the reader is referred to [Buntine (1990)].

3.1.6 *Dempster–Shafer*

The idea of using the Dempster–Shafer theory of evidence [Buchanan and Shortliffe (1984)] for combining classifiers has been suggested in [Shilen (1990)]. This method uses the notion of basic probability assignment defined for a certain class c_i given the instance x:

$$bpa(c_i, x) = 1 - \prod_k \left(1 - \hat{P}_{M_k}(y = c_i \,|x\,)\right) \qquad (3.8)$$

Consequently, the selected class is the one that maximizes the value of the belief function:

$$Bel(c_i, x) = \frac{1}{A} \cdot \frac{bpa(c_i, x)}{1 - bpa(c_i, x)} \qquad (3.9)$$

where A is a normalization factor defined as:

$$A = \sum_{\forall c_i \in dom(y)} \frac{bpa(c_i, x)}{1 - bpa(c_i, x)} + 1 \qquad (3.10)$$

3.1.7 *Vogging*

The idea of behind the vogging approach (Variance Optimized Bagging) is to optimize a linear combination of base-classifiers so as to aggressively reduce variance while attempting to preserve a prescribed accuracy [Derbeko et al. (2002)]. For this purpose, Derbeko et al. implemented the Markowitz Mean-Variance Portfolio Theory that is used for generating low variance portfolios of financial assets.

3.1.8 *Naïve Bayes*

Using Bayes' rule, one can extend the Naïve Bayes idea for combining various classifiers:

$$Class(x) = \operatorname*{argmax}_{\substack{c_j \in dom(y) \\ \hat{P}(y = c_j) > 0}} \hat{P}(y = c_j) \cdot \prod_{k=1} \frac{\hat{P}_{M_k}(y = c_j \,|x\,)}{\hat{P}(y = c_j)} \qquad (3.11)$$

3.1.9 *Entropy Weighting*

The idea in this combining method is to give each classifier a weight that is inversely proportional to the entropy of its classification vector.

$$Class(x) = \underset{c_i \in dom(y)}{\operatorname{argmax}} \sum_{\substack{k:c_i = \operatorname{argmax} \hat{P}_{M_k}(y=c_j|x) \\ c_j \in dom(y)}} E(M_k, x) \qquad (3.12)$$

where:

$$E(M_k, x) = -\sum_{c_j} \hat{P}_{M_k}(y = c_j | x) \log \left(\hat{P}_{M_k}(y = c_j | x) \right) \qquad (3.13)$$

3.1.10 *Density-based Weighting*

If the various classifiers were trained using datasets obtained from different regions of the instance space, it might be useful to weight the classifiers according to the probability of sampling x by classifier M_k, namely:

$$Class(x) = \underset{c_i \in dom(y)}{\operatorname{argmax}} \sum_{\substack{k:c_i = \operatorname{argmax} \hat{P}_{M_k}(y=c_j|x) \\ c_j \in dom(y)}} \hat{P}_{M_k}(x) \qquad (3.14)$$

The estimation of $\hat{P}_{M_k}(x)$ depends on the classifier representation and can not always be estimated.

3.1.11 *DEA Weighting Method*

Recently there has been attempts to use the data envelop analysis (DEA) methodology [Charnes *et al.* (1978)] in order to assign weights to different classifiers [Sohn and Choi (2001)]. These researchers argue that the weights should not be specified according to a single performance measure, but should be based on several performance measures. Because there is a trade-off among the various performance measures, the DEA is employed in order to figure out the set of efficient classifiers. In addition, DEA provides inefficient classifiers with the benchmarking point.

3.1.12   *Logarithmic Opinion Pool*

According to the logarithmic opinion pool [Hansen (2000)] the selection of the preferred class is performed according to:

$$Class(x) = \operatorname*{argmax}_{c_j \in dom(y)} e^{\sum_k \alpha_k \cdot \log(\hat{P}_{M_k}(y=c_j|x))} \tag{3.15}$$

where α_k denotes the weight of the k-th classifier, such that:

$$\alpha_k \geq 0; \sum \alpha_k = 1 \tag{3.16}$$

3.1.13   *Order Statistics*

Order statistics can be used to combine classifiers [Tumer and Ghosh (2000)]. These combiners offer the simplicity of a simple weighted combination method together with the generality of meta-combination methods (see the following section). The robustness of this method is helpful when there are significant variations among classifiers in some part of the instance space.

3.2   Selecting Classification

Recall that ensemble classification can be performed by either fusing the outputs of all members or selecting the output of a single member. In this section we will explore the latter. The premise in this approach is that there is a competent authority that nominates the best classifier for a given instance x. The output of the selected classifier is referred to as the output of the ensemble as a whole.

Very often the input space is partioned into K competence sub-spaces which can have any shape or size. Then for each sub-space we nominate one classifier to be the predictor.

Clustering and classification are both considered fundamental tasks in data–mining. In essence, the difference between clustering and classification lies in the manner knowledge is extracted from data: whereas in classification the knowledge is extracted in a supervised manner based on pre—defined classes, in clustering the knowledge is extracted in an unsupervised way without any guidance from the user.

Decomposition may divide the database horizontally (subsets of rows or tuples) or vertically (subsets of attributes). This section deals with the former, namely tuple decomposition.

Many methods have been developed for partitioning the tuples into subsets. Some of them are aimed at minimizing space and time needed for the classification of a dataset; whereas others attempt to improve accuracy. These methods may be roughly divided according to the manner in which tuples are divided into subsets:

Sample—based tuple decomposition tuples are divided into subsets via sampling. This category includes sampling, a degenerate form of decomposition that decreases complexity but also accuracy [Catlett (1991)], as well as multiple model methods. The latter may be sequential; trying to take advantage of knowledge gained in one iteration, and uses it in the successive one. Such methods include algorithms as windowing [Quinlan (1983)], trying to improve the sample they produce from one iteration to another, and also the boosting algorithm [Schapire (1990)], increasing the probability of selecting instances that are misclassified by the current classifier for constructing the next one, in order to improve accuracy. Sample—based decomposition may also be concurrent, thus enabling parallel learning. Classifiers produced by concurrent methods may be combined using a number of methods, varying from simple voting (e.g. bagging) to more sophisticated meta—classifying methods, such as stacking [Wolpert (1992)], grading [Seewald and Furnkranz (2001)] and arbiter tree [Chan and Stolfo (1993)]. Many multiple model methods were showed to improve accuracy. This accuracy gain may stem from the variation in classifiers, built by the same algorithm, or from the advantages of the sequential process.

Space—based decomposition Tuples are divided into subsets according to their belonging to some part of space. [Kusiak (2000)] describes the notion of "feature value decomposition" in which objects or instances are partitioned into subsets according to the values of selected input attributes. Kusiak also suggests the notion of "decision value decomposition" in which objects are partitioned according to the value of the decision (or more generally, the target attribute). Kusiak does not describe a method for selecting the set of attributes according to which the partition is performed. In fact his work deals only with the decision—making process, and does not offer an automated procedure for space—based decomposition.

A Model Class Selection (MCS) — a system that searches different classification algorithms for different regions in the instance—space is proposed by [Brodley (1995a)]. The MCS system, which can be regarded as implementing an instance—space decomposition strategy, uses dataset characteristics and expert—rules to select one of three possible classification methods (a decision tree, a discriminant function or an instance—based method) for each region in the instance—space. The expert—rules are based on past empirical comparisons of classifier performance, which can be considered as prior knowledge.

In the neural network community, several researchers have examined the decomposition methodology. [Nowlan and Hinton (1991)] examined the Mixture–of–Experts (ME) methodology that decomposes the input space, such that each expert examines a different part of the space. However the subspaces have soft "boundaries", namely subspaces are allowed to overlap. A gating network is responsible for combining the various experts. [Jordan and Jacobs (1994)] have proposed an extension to the basic mixture of experts, known as Hierarchical Mixtures of Experts (HME). This extension decomposes the space into subspaces and then recursively decompose each subspace to subspaces.

Variations of the basic mixture–of–experts method have been developed to accommodate specific domain problems. [Hampshire and Waibel (1992)] and [Peng *et al.* (1996)] have used a specialized modular network called the Meta–pi network to solve the vowel—speaker problem. [Weigend *et al.* (1995)] proposed nonlinear gated experts for time—series while citeOhno–MachadoMusen used a revised modular network for predicting the survival of AIDS patients. [Rahman and Fairhurst (1997)] proposed a new approach for combining multiple experts for improving recognition of handwritten numerals.

NBTree [Kohavi (1996)] is an instance space decomposition method that induces a decision tree and a Naïve Bayes hybrid classifier. Naïve Bayes, which is a classification algorithm based on Bayes' theorem and a Naïve independence assumption, is very efficient in terms of its processing time. To induce an NBTree, the instance space is recursively partitioned according to attributes values. The result of the recursive partitioning is a decision tree whose terminal nodes are Naïve Bayes classifiers. Since subjecting a terminal node to a Naïve Bayes classifier means that the hybrid classifier may classify two instances from a single hyper—rectangle region into distinct classes, the NBTree is more flexible than a pure decision tree. In order to decide when to stop the growth of the tree, NBTree compares two alternatives in terms of error estimation — partitioning into a hyper—rectangle regions and inducing a single Naïve Bayes classifier. The error estimation is calculated by cross—validation, which significantly increases the overall processing time. Although NBTree applies a Naïve Bayes classifier to decision tree terminal nodes, classification algorithms other than Naïve Bayes are also applicable. However, the cross—validation estimations make the NBTree hybrid computationally expensive for more time—consuming algorithms such as neural networks.

NBTree uses a simple stopping criterion according to which a split is not considered when the dataset consists of 30 instances or less. Splitting too few instances will not affect the final accuracy much yet will lead, on the other hand, to a complex composite classifier. Moreover, since each sub classifier is required to generalize instances in its region, it must be trained on samples of sufficient size. [Kohavi (1996)] suggested a new splitting criterion which is to select the attribute with the highest utility. Kohavi defined utility as the 5—fold cross—validation accuracy estimate of using a Naïve Bayes algorithm for classifying regions generated by a split. The regions are partitions of the initial subspace according to a particular attribute values.

Although different researchers have addressed the issue of instance space decomposition, there is no research that suggests an automatic procedure for mutually exclusive instance space decompositions, which can be employed for any given classification algorithm and in a computationally efficient way. We present an algorithm for space decomposition, which exploits the K—means clustering algorithm. It is aimed at reducing the error rate comparing to the simple classifier embedded in it, while keeping the comprehensibility level.

3.2.1 *Partitioning the Instance Space*

This section presents a decomposition method that partitions the instance space using the K—means algorithm and then employs an induction algorithm on each cluster. Because space decomposition is not necessarily suitable to any given dataset and in some cases it might reduce the classification accuracy, we suggest a homogeneity index that measures the initial reduction in sum of square errors resulting from the clustering procedure. Consequently the decomposition method is executed only if the homogeneity index obtained a certain threshold value. Additionally the proposed procedure ensure that there is suffices number of instances in each cluster for inducing a classifier. An empirical study conducted shows that the proposed method can lead to a significant increase of classification accuracy especially in numeric datasets.

One of the main issues arising when trying to address the problem formulated in the last section concerns the question of what sort of instance space division should be taken in order to achieve as high accuracy as possible. One may come up with quite a few ways for dividing the instance space, varying from using one attribute at a time (similarly to decision tree construction) to the use of different combinations of attribute values.

Inspired by the idea that similar instances should be assigned to the same subspace, it lead us towards using some clustering method as a possible tool for detecting populations. That is since "clustering is the grouping of similar objects" [Hartigan (1975)]. We choose to define the similarity of unlabeled data via the distance metric. In particular, the metric used will be the Euclidean metric for continuous attributes, involving simple matching for nominal ones (very similar to the similarity measure used by [Haung (1998)] in the K—prototypes algorithm, except for the fact that there is no special cluster—dependent weight for the categorical attributes). The reason for this particular metric chosen lies in the clustering method we prefer for this work, namely the K—means algorithm.

3.2.1.1 *The K–Means Algorithm as a Decomposition Tool*

The K—means algorithm is one of the simplest and most commonly used clustering algorithms. It is a partitional algorithm, heuristically attempting to minimize the sum of squared errors:

$$SSE = \sum_{k=1}^{K} \sum_{i=1}^{N_k} \|x_i - \mu_k\|^2 \qquad (3.17)$$

where N_k is the number of instances belonging to cluster k and μ_k is the mean of k'th cluster, calculated as the mean of all the instances belonging to that cluster:

$$\mu_{k,i} = \frac{1}{N_k} \sum_{q=1}^{N_k} x_{q,i} \forall i \qquad (3.18)$$

Figure 3.1 presents the pseudo—code of the K—means algorithm. The algorithm starts with an initial set of cluster centers, chosen at random or according to some heuristic procedure. In each iteration, each instance is assigned to its nearest cluster center according to the Euclidean distance between the two. Then the cluster centers are re—calculated.

A number of convergence conditions are possible. For example, the search may stop when the partitioning error is not reduced by the relocation of the centers. This indicates that the present partition is locally optimal. Other stopping criteria can be used also such as exceeding a pre—defined number of iterations.

K-Mean Clustering (S, K)

S - Instances Set

K - Number of Clusters

Randomly initialize K cluster centers.

WHILE termination condition is not satisfied {

    Assign instances to the closest cluster center.

    Update cluster centers using the instances assignment

}

Fig. 3.1   K-means algorithm.

The K—means algorithm may be viewed as a gradient—decent procedure, which begins with an initial set of K cluster—centers and iteratively updates it so as to decrease the error function. The algorithm starts with an initial set of cluster centers, chosen at random or according to some heuristic. In each iteration, each instance is assigned to its nearest cluster center according to the Euclidean distance between the two. Then the cluster centers are re—calculated.

A number of convergence conditions are possible, including no reduction in error as a result of the relocation of centers, no (or minimal) reassignment of instances to new cluster centers, or exceeding a pre—defined number of iterations. A rigorous proof of the finite convergence of the K—means type algorithms is given in [Selim and Ismail (1984)]. The complexity of T iterations of the K—means algorithm performed on a sample size of m instances, each characterized by N attributes is: $O(T * K * m * N)$.

For T iterations of the K—means algorithm performed on a dataset containing m instances each has n attributes, its complexity may be calculated as: $O(T * K * m * n)$. This linear complexity with respect to m is one of the reasons for the popularity of K—means: Even if the number of instances is substantially large (which often is the case nowadays) — this algorithm is computationally attractive. Thus, K—means has an advantage in comparison to other clustering methods (e.g. hierarchical clustering methods), which have non—linear complexity with respect to the number of instances.

Other reasons for the algorithm's popularity are its ease of interpretation, simplicity of implementation, speed of convergence and adaptability to sparse data [Dhillon and Modha (2001)].

Having intended to use a clustering algorithm as a means for partitioning the dataset, and taking into account the availability, linear complexity and high understandability of the K—means algorithm, we choose to integrate this specific clustering method in our algorithm.

The K—means algorithm may be considered as a simplification of the expectation maximization algorithm [Dempster *et al.* (1977)]. This is a density based clustering algorithm used for identifying the parameters of different distributions from which the data objects are assumed to be drawn. In the case of K—means, the objects are assumed to be drawn from a mixture of K multivariate normal distributions, sharing the same known variance whereas the mean vectors of the K distributions are unknown [Estivill-Castro (2000)]. When employing the K—means on the unlabeled data, this underlying assumption of the algorithm may be written as:

$$x \sim N(\mu_k, \sigma^2) \forall k = 1, 2, \ldots, K, x \in C_k \tag{3.19}$$

According to Bayes' theorem:

$$p(y = c_j^* \,|x\,) = \frac{p(y = c_j^*, x)}{p(x)} \tag{3.20}$$

Since $p(x)$ depends on the distribution from which the unlabeled instances are drawn and since it is plausible to assume that different clusters has different distributions, it implies that $p(y = c_j^* \,|x\,)$ is distributed differently on different clusters. The latter distribution has a direct influence on the predicted value of the target attribute, since:

$$\hat{y}(x) = \underset{c_j^* \in dom(y)}{\arg\max}\, p\left(y = c_j^* \,|x\,\right) \tag{3.21}$$

This supports the idea of using clustering algorithm.

3.2.1.2 *Determining the Number of Subsets*

In order to proceed with the decomposition of unlabeled data, a significant parameter should be at hand — the number of subsets, or in our case, clusters, existing in the data.

The K—means algorithm requires this parameter as input, and is affected by its value. Various heuristics attempt to find an optimal number of clusters most of them refer to inter—cluster distance or intra—cluster similarity. Nevertheless in this case as we know the actual class of each instance, we suggest using the mutual information criterion for clustering [Strehl *et al.* (2000)]). The criterion value for m instances clustered using $C = \{C_1, \ldots, C_g\}$ and referring to the target attribute y whose domain is $dom(y) = \{c_1, \ldots, c_k\}$ is defined as follows:

$$C = \frac{2}{m} \sum_{l=1}^{g} \sum_{h=1}^{k} m_{l,h} \log_{g \cdot k} \left(\frac{m_{l,h} \cdot m}{m_{.,l} \cdot m_{l,.}} \right) \tag{3.22}$$

where $m_{l,h}$ indicate the number of instances that are in cluster C_l and also in class c_h. $m_{.,h}$ denotes the total number of instances in the class c_h. Similarly $m_{l,.}$ indicates the number of instances in cluster C_l.

3.2.1.3 *The Basic K–Classifier Algorithm*

The basic K—classifier algorithm employs the K—means algorithm for the purpose of space decomposition and uses mutual information criterion for clustering for determining the number of clusters. The algorithm follows the following steps:

Step 1 Apply the K—means algorithm to the training set S using $K = 2, 3, \ldots K_{max}$

Step 2 Compute the mutual information criterion for clustering for $K = 2, 3, \ldots, K_{max}$ and choose the optimal number of clusters K^*.

Step 3 Produce K classifiers of the induction algorithm I, each produced on the training data belonging to a subset k of the instance space. A decomposition of the space is defined as follows: $B_k = \{x \in X : k = \arg\min \|x - \mu_k\|\}$ $k = 1, 2, \ldots, K^*$ and therefore the classifier constructed will be: $I(x \in S \cap B_k)$ $k = 1, 2, \ldots, K^*$

New instances are classified by the K—classifier as follows:

- The instance is assigned to the cluster closest to it: B_k : $k = \arg \min \|x - \mu_k\|$.
- The classifier induced using B_k is employed for assigning a class to the instance.

We analyze the extent to which the conditions surrounding the basic K—classifier may lead to its success or failure. This is done using three representative classification algorithms: C4.5, Neural network and Naïve Bayes. These algorithms, denoted by "DT", "ANN" and "NB" respectively, are employed on eight databases from the UCI repository, once in their basic form and once combined with the K—classifier. The classification error rate, resulting from the decomposition, is measured and compared to that achieved by the basic algorithm using McNemar's test [Dietterich (1998)]. The maximum number of clusters is set to a sufficiently large number (25).

These experiments are executed 5 times for each database and each classifying algorithm, in order to reduce the variability resulting from the random choice of training set in McNemar's test.

In order to analyze the causes for the K—classifier's success/failure, a Meta dataset has been constructed. This dataset contains a tuple for each experiment on each database with each classifying algorithm. Its attributes correspond to the characteristics of the experiment:

Record—attribute ratio Calculated as training set size divided by the attribute set size.

Initial PRE the reduction in the SSE, resulting from partitioning the dataset from one cluster (the non partitioned form) to two. This characteristic was chosen since we suspect it indicates whether the data set should be partitioned at all.

Induction method the induction algorithm employed on the database.

In order to analyze the reduction in error rate as a function of method and dataset characteristics, a meta—classifier is constructed. The inducer employed for this purpose is the C4.5 algorithm.

As for the target attribute it represents the accuracy performance of the basic K—classifier algorithm relatively to the appropriate accuracy performance of the inducer employed in the base form. The target attribute can have one of the following values: non—significant decrease/increase of up to ten percent ("small ns dec/inc"), non—significant decrease/increase of ten percent or more ("large ns dec/inc"), significant decrease/increase of up to ten percent ("small s dec/inc"), significant decrease/increase of ten percent or more ("large s dec/inc"), and a decrease rate of 0 percent ("no change"). The resultant decision tree is presented in Figure 3.2.

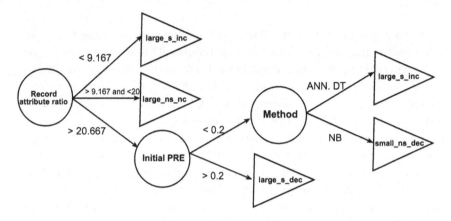

Fig. 3.2   A decision tree describing the change in error achieved by the K—classifier.

As may be learned from the tree, the two attributes that determine whether or not the K—classifier would achieve a significant decrease in error rate are the record—attribute ratio and the initial PRE. A significant decrease in error rate may occur when the former characteristic exceeds 20.67 and the latter exceeds 0.2.

This result also answers the question: should there always be a recommended partition? This question may be viewed as a preliminary check of the dataset, aiming at discovering whether or not it requires space decomposition.

When the record—attribute ratio is smaller than 20.67 or equals it, the result will be a significant increase in the error rate, or at the very least — a non—significant increase. Therefore, it may be concluded that datasets containing a small number of records compared to their number of attributes should not be partitioned using the K—classifier algorithm.

Another conclusion that may be drawn from this stage is that the K—classifier algorithm works better on integer or continuous—valued attributes. Though the algorithm did not significantly decrease error on all databases of such values, the ones on which error decreased significantly all contained integer attributes, continuous—valued attributes or some combination of these two kinds.

3.2.1.4 *The Heterogeneity Detecting K–Classifier (HDK–Classifier)*

The analysis of error reduction rate provides the basic K—classifier with the missing link regarding when clustering should be used. As we suspected, decomposition does not always yield an accuracy gain, and may deteriorate it on many occasions. Such results may derive from the homogeneity, or lack of heterogeneity of the dataset: there are no distinct clusters or populations in the dataset, and therefore it should not be partitioned.

The mutual information criterion, used in the basic K—classifier, does not examine whether heterogeneity exists. It simply assumes it exits and aims at finding the number of populations in the data, given that the dataset is indeed composed of different populations.

Should we detect non—heterogeneous datasets, there is no need for decomposing them, since their error will not decrease. The current K—classifier, increasing the running time complexity compared to the basic learning algorithm, only yields worse results on such datasets. In light of this we refine the basic K—classifier, and add another step, to be taken first. In this step the K—means is employed for $K = 1$ and $K = 2$, and it is checked whether the resultant PRE is larger than 0.2. If so, the rest of the K—classifier stages follow. If not, it is assumed that there is no use in decomposition, so the inducer in its base form is employed on the entire dataset. Thus, the algorithm maintains the accuracy of non—heterogeneous datasets at the expense of an additional complexity that is much smaller compared to the basic K—classifier.

3.2.1.5 *Running–Time Complexity*

The offered training algorithm requires the following computations:

- During the stages of determining the best number of clusters, the K—means algorithm is run Kmax-1 times. That leads to a complexity of $O(T * K_{max}^2 * n * m)$.
- Computation of the PRE's value for $K = 1$ and $K = 2$ is of $O(n * m)$ complexity and is therefore negligible.
- Constructing a classifier on each of the K^* partitions requires at most $O(K_{max} * G_I(m, n))$ where G_I is the classifier's training complexity time. For instance, when employing the decision tree algorithm, the time complexity of this stage will be at most $O(K_{max} * m\sqrt{l})$ where the number of leaves of the decision tree is l.

In light of the above analysis the total running—time complexity of the training algorithm is $O(T * K_{max}^2 * n * m + K_{max} * G_I(m, n))$. In the case of decision trees classifiers, for instance, the time—complexity would be: $O(T * K_{max}^2 * n * m + K_{max} * m\sqrt{l})$.

3.3   Mixture of Experts and Meta Learning

Meta-learning is a process of learning from learners (classifiers). The training of a meta-classifier is composed of two or more stages, rather than one stage, as with standard learners. In order to induce a meta classifier, first the base classifiers are trained (stage one), and then the Meta classifier (second stage). In the prediction phase, base classifiers will output their classifications, and then the Meta-classifier(s) will make the final classification (as a function of the base classifiers). Meta-learning methods are best suited for cases in which certain classifiers consistently correctly classify, or consistently misclassify, certain instances.

The following sections describe the most well-known meta-combination methods.

3.3.1   *Stacking*

Stacking is probably the most-popular meta-learning technique [Wolpert (1992)]. By using a meta-learner, this method tries to induce which classifiers are reliable and which are not. Stacking is usually employed to combine models built by different inducers. The idea is to create a meta-dataset containing a tuple for each tuple in the original dataset. However, instead of using the original input attributes, it uses the predicted classifications by the classifiers as the input attributes. The target attribute remains as in the original training set. A test instance is first classified by each of the base classifiers. These classifications are fed into a meta-level training set from which a meta-classifier is produced. This classifier combines the different predictions into a final one. It is recommended that the original dataset should be partitioned into two subsets. The first subset is reserved to form the meta-dataset and the second subset is used to build the base-level classifiers. Consequently the meta-classifier predications reflect the true performance of base-level learning algorithms. Stacking performance can be improved by using output probabilities for every class label from the base-level classifiers. In such cases, the number of input attributes in the meta-dataset is multiplied by the number of classes.

It is recommended that the original dataset should be partitioned into two subsets. The first subset is reserved to form the meta-dataset and the second subset is used to build the base-level classifiers. Consequently the meta-classifier predications reflect the true performance of base-level learning algorithms. Stacking performance can be improved by using output probabilities for every class label from the base-level classifiers. It has been shown that with stacking the ensemble performs (at best) comparably to selecting the best classifier from the ensemble by cross validation (Dzeroski and Zenko, 2004).

It has been shown that with stacking the ensemble performs (at best) comparably to selecting the best classifier from the ensemble by cross validation [Džeroski and Ženko (2004)]. In order to improve the existing stacking approach, they employed a new multi-response model tree to learn at the meta-level and empirically showed that it performs better than existing stacking approaches and better than selecting the best classifier by cross-validation.

There are many variants of the basic Stacking algorithm [Wolpert and Macready (1996)]. The most useful Stacking scheme is specified in [Ting and Witten (1999)]. The meta-database is composed of the posteriori class probabilities of each classifier. it has been shown that this schema in combination with multi-response linear regression as a meta-learner gives the best results.

Džeroski and ženko (2004) demonstrated that for this schema to work better than just selecting the best classifier, it is required the use of a meta-classifier based on multi-response trees. Seewald (2002A) showed that from a choice of seven classifiers it was possible for a Stacking scheme using four of the classifiers that were considered as belonging to different classifier types, to perform equally as well as all seven classifiers. There has been considerably less intention given to the area of heterogeneity and Stacking in the area of regression problems.

StackingC is a variation of the simple Stacking method. In empirical tests Stacking showed significant performance degradation for multi-class datasets. StackingC was designed to address this problem. In StackingC, each base classifier outputs only one class probability prediction (Seewald, 2003). Each base classifier is trained and tested upon one particular class while stacking output probabilities for all classes and from all component classifiers.

Seewald (2003) has shown that all ensemble learning systems, including StackingC (Seewald, 2002B), Grading (Seewald and Fuernkranz, 2001) and even Bagging (Breiman, 1996) can be simulated by Stacking (Wolpert, 1992). To do this they give functionally equivalent definitions of most schemes as Meta-classifiers for Stacking. Džeroski and ženko (2004) indicated that the combination of SCANN (Merz, 1999), which is a variant of Stacking, and MDT (Ting and Witten, 1999) plus selecting the best base classifier using cross validation seems to perform at about the same level as Stacking with Multi-linear Response (MLR).

Seewald (2003) presented strong empirical evidence that Stacking in the extension proposed by Ting and Witten (1999) performs worse on multi-class than on two-class datasets, for all but one meta-learner he investigated. The explanation given was that when the dataset has a higher number of classes, the dimensionality of the meta-level data is proportionally increased. This higher dimensionality makes it harder for meta-learners to induce good models, since there are more features to be considered. The increased dimensionality has two more drawbacks. First, it increases the training time of the Meta classifier; in many inducers this problem is acute. Second, it also increases the amount of memory which is used in the process of training. This may lead to insufficient resources, and therefore may limit the number of training cases (instances) from which an inducer may learn, thus damaging the accuracy of the ensemble.

During the learning phase of StackingC it is essential to use one-against-all class binarization and regression learners for each class model. This class binarization is believed to be a problematic method especially when class distribution is highly non-symmetric. It has been illustrated (Frnkranz, 2002) that handling many classes is a major problem for the one-against-all binarization technique, possibly because the resulting binary learning problems increasingly skewed class distributions. An alternative to one-against-all class binarization is the one-against-one binarization in which the basic idea is to convert a multiple class problem into a series of two-class problems by training one classifier for each pair of classes, using only training examples of these two classes and ignoring all others. A new example is classified by submitting it to each of the $\frac{k(k-1)}{2}$ binary classifiers, and combining their predictions. We have found in our preliminary experiments that this binarization method yields noticeably poor accuracy results when the number of classes in the problem increases. Later, after performing a much wider and broader experiment on StackingC in conjunction with the one-against-one binarization method, we came to this same conclusion. An explanation might be that, as the number of classes in a problem increases, the greater is the chance that any of the $\frac{k(k-1)}{2}$ base classifiers will give a wrong prediction. There are two reasons for this. First, when predicting the class of an instance, only out of $\frac{k(k-1)}{2}$ classifiers may predict correctly. This is because only $k-1$ classifiers were trained on any specific class.

The second reason is that in one-against-one binarization we use only instances of two classes – the instances of each one of the pair classes, while in one-against-all we use all instances, and thus the number of training instances for each base classifier in one-against-one binarization is much smaller than in the one-against-all binarization method. Thus using the one-against-one binarization method may yield inferior base classifier.

StackingC improves on Stacking in terms of significant accuracy differences, accuracy ratios, and runtime. These improvements are more evident for multi-class datasets and have a tendency to become more pronounced as the number of classes increases. StackingC also resolves the weakness of Stacking in the extension proposed by Ting and Witten (1999) and offers a balanced performance on two-class and multi-class datasets.

The SCANN (Stacking, Correspondence Analysis and Nearest Neighbor) combining method [Merz (1999)] uses the strategies of stacking and correspondence analysis. Correspondence analysis is a method for geometrically modelling the relationship between the rows and columns of a matrix whose entries are categorical. In this context Correspondence Analysis is used to explore the relationship between the training examples and their classification by a collection of classifiers.

A nearest neighbor method is then applied to classify unseen examples. Here, each possible class is assigned coordinates in the space derived by Correspondence Analysis. Unclassified examples are mapped into the new space, and the class label corresponding to the closest class point is assigned to the example.

3.3.2 *Arbiter Trees*

According to Chan and Stolfo's approach [Chan and Stolfo (1993)], an arbiter tree is built in a bottom-up fashion . Initially, the training set is randomly partitioned into k disjoint subsets. The arbiter is induced from a pair of classifiers and recursively a new arbiter is induced from the output of two arbiters. Consequently for k classifiers, there are $\log_2(k)$ levels in the generated arbiter tree.

The creation of the arbiter is performed as follows. For each pair of classifiers, the union of their training dataset is classified by the two classifiers. A selection rule compares the classifications of the two classifiers and selects instances from the union set to form the training set for the arbiter. The arbiter is induced from this set with the same learning algorithm used in the base level. The purpose of the arbiter is to provide an alternate classification when the base classifiers present diverse classifications. This arbiter, together with an arbitration rule, decides on a final classification outcome, based upon the base predictions. Figure 3.3 shows how the final classification is selected based on the classification of two base classifiers and a single arbiter.

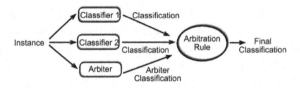

Fig. 3.3   A prediction from two base classifiers and a single arbiter.

The process of forming the union of data subsets; classifying it using a pair of arbiter trees; comparing the classifications; forming a training set; training the arbiter; and picking one of the predictions, is recursively performed until the root arbiter is formed. Figure 3.4 illustrate an arbiter tree created for $k = 4$. $T_1 - T_4$ are the initial four training datasets from which four classifiers $M_1 - M_4$ are generated concurrently. T_{12} and T_{34} are the training sets generated by the rule selection from which arbiters are produced. A_{12} and A_{34} are the two arbiters. Similarly, T_{14} and A_{14} (root arbiter) are generated and the arbiter tree is completed.

There are several schemes for arbiter trees; each is characterized by a different selection rule. Here are three versions of selection rules:

- Only instances with classifications that disagree are chosen (group 1).
- Like group 1 defined above, plus instances where their classifications agree but are incorrect (group 2).
- Like groups 1 and 2 defined above, plus instances that have the same correct classifications (group 3).

Of the two versions of arbitration rules that have been implemented, each corresponds to the selection rule used for generating the training data at

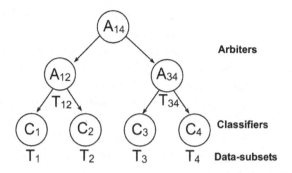

Fig. 3.4   Sample arbiter tree.

that level:

- For selection rule 1 and 2, a final classification is made by a majority vote of the classifications of the two lower levels and the arbiter's own classification, with preference given to the latter.
- For selection rule 3, if the classifications of the two lower levels are not equal, the classification made by the sub-arbiter based on the first group is chosen. In case this is not true and the classification of the sub-arbiter constructed on the third group equals those of the lower levels, then this is the chosen classification. In any other case, the classification of the sub-arbiter constructed on the second group is chosen. In fact it is possible to achieve the same accuracy level as in the single mode applied to the entire dataset but with less time and memory requirements [Chan and Stolfo (1993)]. More specifically it has been shown that this meta-learning strategy required only around 30% of the memory used by the single model case. This last fact, combined with the independent nature of the various learning processes, make this method robust and effective for massive amounts of data. Nevertheless, the accuracy level depends on several factors such as the distribution of the data among the subsets and the pairing scheme of learned classifiers and arbiters in each level. The decision regarding any of these issues may influence performance, but the optimal decisions are not necessarily known in advance, nor initially set by the algorithm.

3.3.3 *Combiner Trees*

The way combiner trees are generated is very similar to arbiter trees. Both are trained bottom-up. However, a combiner, instead of an arbiter, is placed in each non-leaf node of a combiner tree [Chan and Stolfo (1997)]. In the combiner strategy, the classifications of the learned base classifiers form the basis of the meta-learner's training set. A composition rule determines the content of training examples from which a combiner (meta-classifier) will be generated. In classifying an instance, the base classifiers first generate their classifications and based on the composition rule, a new instance is generated. The aim of this strategy is to combine the classifications from the base classifiers by learning the relationship between these classifications and the correct classification. Figure 3.5 illustrates the result obtained from two base classifiers and a single combiner.

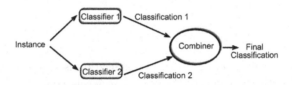

Fig. 3.5 A prediction from two base classifiers and a single combiner.

Two schemes for composition rules were proposed. The first one is the stacking scheme. The second is like stacking with the addition of the instance input attributes. It has been shown that the stacking scheme per se does not perform as well as the second scheme [Chan and Stolfo (1995)]. Although there is information loss due to data partitioning, combiner trees can sustain the accuracy level achieved by a single classifier. In a few cases, the single classifier's accuracy was consistently exceeded.

3.3.4 *Grading*

This technique uses "graded" classifications as meta-level classes [Seewald and Furnkranz (2001)]. The term "graded" is used in the sense of classifications that have been marked as correct or incorrect. The method transforms the classification made by the k different classifiers into k training sets by using the instances k times and attaching them to a new binary class in each occurrence. This class indicates whether the k-th classifier yielded a correct or incorrect classification, compared to the real class of the instance.

For each base classifier, one meta-classifier is learned whose task is to classify when the base classifier misclassifies. At classification time, each base classifier classifies the unlabeled instance. The final classification is derived from the classifications of those base classifiers that are classified to be correct by the meta-classification schemes. In case several base classifiers with different classification results are classified as correct, voting, or a combination considering the confidence estimates of the base classifiers, is performed. Grading may be considered as a generalization of cross-validation selection [Schaffer (1993)], which divides the training data into k subsets, builds $k - 1$ classifiers by dropping one subset at a time and then uses it to find a misclassification rate. Finally, the procedure simply chooses the classifier corresponding to the subset with the smallest misclassification. Grading tries to make this decision separately for each and every instance by using only those classifiers that are predicted to classify that instance correctly. The main difference between grading and combiners (or stacking) is that the former does not change the instance attributes by replacing them with class predictions or class probabilities (or adding them to it). Instead it modifies the class values. Furthermore, in grading several sets of meta-data are created, one for each base classifier. Several meta-level classifiers are learned from those sets.

The main difference between grading and arbiters is that arbiters use information about the disagreements of classifiers for selecting a training set; grading uses disagreement with the target function to produce a new training set.

3.3.5 *Gating Network*

Figure 3.6 illustrates an n-expert structure. Each expert outputs the conditional probability of the target attribute given the input instance. A gating network is responsible for combining the various experts by assigning a weight to each network. These weights are not constant but are functions of the input instance x. The gating network selects one or a few experts (classifiers) which appear to have the most appropriate class distribution for the example. In fact each expert specializes on a small portion of the input space.

An extension to the basic mixture of experts, known as hierarchical mixtures of experts (HME), has been proposed in [Jordan and Jacobs (1994)].

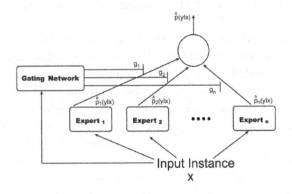

Fig. 3.6   Illustration of n-expert structure.

This extension decomposes the space into sub-spaces, and then recursively decomposes each sub-space into sub-spaces.

Variations of the basic mixtures of experts methods have been developed to accommodate specific domain problems. A specialized modular networks called the Meta-p_i network has been used to solve the vowel-speaker problem [Hampshire and Waibel (1992); Peng *et al.* (1996)]. There have been other extensions to the ME such as nonlinear gated experts for time-series [Weigend *et al.* (1995)]; revised modular network for predicting the survival of AIDS patients [Ohno-Machado and Musen (1997)]; and a new approach for combining multiple experts for improving handwritten numeral recognition [Rahman and Fairhurst (1997)].

Some of the weighting methods are trainable. Lin *et al.* (2005) propose to use genetic algorithms in attempt to find the optimal weights. They describe two different combinatorial schemes to improve the performance of handwritten Chinese character recognition: the accuracy rate of the first candidate class and the accuracy rate of top ten candidate classes. Their extensive study show that this new approach can significantly improve the accuracy performance.

Reinforcement learning (RL) has been used to adaptively combine the base classifiers [Dimitrakakis (2005)]. The ensemble consists of a controlling agent that selects which base classifiers are used to classify a particular instance. The controlling agent learns to make decisions so that classification error is minimized. The agent is trained through a Q-learning inspired technique. The usage of reinforcement learning improves results when there are many base classifiers.

Chapter 4

Ensemble Diversity

4.1 Overview

Ensemble methods are very effective, mainly due to the phenomenon that various types of classifiers have different "inductive biases" [Mitchell (1997)]. In order to make the ensemble more effective, there should be some sort of diversity between the classifiers [Kuncheva (2005b)]. Diversity may be obtained through different presentations of the input data, as in bagging, variations in learner design, or by adding a penalty to the outputs to encourage diversity.

Indeed, ensemble methods can effectively make use of such diversity to reduce the variance-error [Tumer and Ghosh (1996); Ali and Pazzani (1996)] without increasing the bias-error. In certain situations, an ensemble can also reduce bias-error, as shown by the theory of large margin classifiers [Bartlett and Shawe-Taylor (1998)]. In an ensemble, the combination of the output of several classifiers is only useful if they disagree about some inputs [Tumer and Ghosh (1996)].

Creating an ensemble in which each classifier is as different as possible while still being consistent with the training set is theoretically known to be an important feature for obtaining improved ensemble performance [Krogh and Vedelsby (1995)]. According to [Hu (2001)], diversified classifiers lead to uncorrelated errors, which in turn improve classification accuracy .

Brown et al. [Brown et al. (2005)] indicate that for classification tasks the concept of "diversity" is still an ill-defined concept. Nevertheless it is believed to be closely related to the statistical concept of correlation. Diversity is obtained when the misclassification events of the base classifiers are not correlated. Several means can be used to reach this goal: different presentations of the input data, variations in learner design, or by adding a penalty to the outputs to encourage diversity.

In the regression context, the bias-variance-covariance decomposition has been suggested to explain why and how diversity between individual models contribute toward overall ensemble accuracy. Nevertheless, in the classification context, there is no complete and agreed upon theory [Brown *et al.* (2005)]. More specifically, there is no simple analogue of variance-covariance decomposition for the zero-one loss function. Instead, there are several ways to define this decomposition. Each way has its own assumptions.

Sharkey [Sharkey (1999)] suggested a taxonomy of methods for creating diversity in ensembles of neural networks. More specifically, Sharkey's taxonomy refers to four different aspects: the initial weights; the training data used; the architecture of the networks; and the training algorithm used.

Brown *et al.* [Brown *et al.* (2005)] suggest a different taxonomy which consists of the following branches: varying the starting points within the hypothesis space; varying the set of hypotheses that are accessible by the ensemble members (for instance by manipulating the training set); and varying the way each member traverses the space.

In this chapter we suggest the following taxonomy. Note however that the components of this taxonomy are not mutually exclusive, namely, there are a few algorithms which combine two of them.

(1) Manipulating the Inducer – We manipulate the way in which the base inducer is used. More specifically each ensemble member is trained with an inducer that is differently manipulated.

(2) Manipulating the Training Sample – We vary the input that is used by the inducer for training. Each member is trained from a different training set.

(3) Changing the target attribute representation – Each classifier in the ensemble solve a different target concept.

(4) Partitioning the search space – Each member is trained on a different search subspace.

(5) Hybridization – Diversity is obtained by using various base inducers or ensemble strategies.

4.2 Manipulating the Inducer

A simple method for gaining diversity is to manipulate the inducer used for creating the classifiers. Below we survey several strategies to gain this diversity.

4.2.1   *Manipulation of the Inducer's Parameters*

The base inducer usually can be controlled by a set of parameters. For example, the well known decision tree inducer C4.5 has the confidence level parameter that greatly affect learning. Drucker [Drucker (2002)] examine the effect of early pruning of decision trees on the performance of the entire ensemble. When an algorithm (such as decision tree) is used as a single strong learner, then certain aspects should be taken into consideration. But when the same algorithm is used as a weak learner then other aspects should be taken into consideration.

In the neural network community, there were several attempts to gain diversity by using different number of nodes [Partridge and Yates (1996); Yates and Partridge (1996)]. Nevertheless, these researches concludes that variation in numbers of hidden nodes is not effective method of creating diversity in neural network ensembles. Nevertheless the CNNE algorithm [Islam *et al.* (2003)] which simultaneously determines the ensemble size along with the number of hidden nodes in individual NNs, has shown encouraging results.

Another effective approach for ANNs is to use several network topologies. For instance the Addemup algorithm [Opitz and Shavlik (1996)] uses genetic algorithm to select the network topologies composing the ensemble. Addemup trains with standard backpropagation, then selects groups of networks with a good error diversity according to the measurement of diversity.

4.2.2   *Starting Point in Hypothesis Space*

Some inducers can gain diversity by starting the search in the Hypothesis Space from different points. For example the simplest way to manipulate the back-propagation inducer is to assign different initial weights to the network [Kolen and Pollack (1991)]. Experimental study indicate that the resulting networks differed in the number of cycles in which they took to converge upon a solution, and in whether they converged at all. While it is very simple way to gain diversity, it is now generally accepted that it is not sufficient for achieving good diversity [Brown *et al.* (2005)].

4.2.3   *Hypothesis Space Traversal*

These techniques alter the way the inducer traverses the space, thereby leading different classifiers to converge to different hypotheses [Brown *et al.*

(2005)]. We differentiate between two techniques for manipulating the space traversal for gaining diversity: Random and Collective-Performance.

Random-based strategy

The idea in this case is to "inject randomness" into the inducers in order to increase the independence among the ensemble's members. Ali and Pazzani [Ali and Pazzani (1996)] propose to change the rule learning HYDRA algorithm in the following way: Instead of selecting the best attribute at each stage (using, for instance, an information gain measure), the attribute is selected randomly such that its probability of being selected is proportional to its measured value. A similar idea has been implemented for C4.5 decision trees [Dietterich (2000a)]. Instead of selecting the best attribute in each stage, it selects randomly (with equal probability) an attribute from the set of the best 20 attributes.

Collective-Performance-based strategy

In this case the evaluation function used in the induction of each member is extended to include a penalty term that encourages diversity. The most studied penalty method is the Negative Correlation Learning [Brown and Wyatt (2003); Rosen (1996)]. The idea of negative correlation learning is to encourage different individual classifiers in the ensemble to represent different subspaces of the problem. While simultaneously creating the classifiers, the classifiers may interact with each other in order to specialize (for instance by using a correlation penalty term in the error function to encourage such specialization).

4.3   Manipulating the Training Samples

In this method, each classifier is trained on a different variation or subset of the original dataset. This method is useful for inducers whose variance-error factor is relatively large (such as decision trees and neural networks). That is to say, small changes in the training set may cause a major change in the obtained classifier. This category contains procedures such as bagging, boosting and cross-validated committees.

4.3.1   *Resampling*

The distribution of tuples among the different classifier could be random as in the bagging algorithm or in the arbiter trees. Other methods distribute the tuples based on the class distribution such that the class distribution in each subset is approximately the same as that in the entire dataset. It has been shown that proportional distribution as used in combiner trees [Chan and Stolfo (1995)] can achieve higher accuracy than random distribution.

Instead of perform sampling with replacement, some methods (like AdaBoost or Wagging) manipulate the weights that are attached to each instance in the training set. The base inducer should be capable to take these weights into account. Recently a novel framework was proposed in which each instance contributes to the committee formation with a fixed weight, while contributing with different individual weights to the derivation of the different constituent classifiers [Christensen *et al.* (2004)]. This approach encourages model diversity without biasing the ensemble inadvertently towards any particular instance.

The bagging using diversity (BUD) algorithm [Tang *et al.* (2006)] elaborates on the bagging algorithm by considering the diversity among the base classifiers in order to achieve better results. It operates on the assumption that the more diverse the classifiers are from each other, the better results that are achieved the classifiers are combined. The algorithm generates a set of base-classifiers from the training instances and then selects a subset of the generated base-classifiers by iteratively applying different diversity-measures on the current ensemble and the potential base-classifier which is to be added to the ensemble.

Figure 4.1 presents the pseudocode of the BUD algorithm. First, a simple bagging ensemble with T base classifiers is created. The classifier with the smallest training error constitutes the initial output ensemble M'. Using either "disagreement", the "Kohavi-Wolpert variance" or "generalized diversity", additional $T/2 - 1$ bases classifiers are added to M'. Classification of new instances is performed using distribution summation by taking into account the output of the base classifiers included in M'. Figures 4.2 to 4.6 present the pseudo code for calculating the diversity measures.

4.3.2 *Creation*

The DECORATE algorithm [Melville and Mooney (2003)] is a dependent approach in which the ensemble is generated iteratively, learning a classifier at each iteration and adding it to the current ensemble. The first member is created by using the base induction algorithm on the original training set.

Bagging using diversity.

Require: I (a base inducer), T (number of iterations), S (the original training set), D (diversity measure).

1: **for** $t = 1$ to T **do**
2:     Create a dataset S' of size of S, created using random sampling with replacement.
3:     $M_t = I(S')$
4: **end for**
5: $M' = \underset{t \in [1,T]}{\arg\min} \sum_{x : M_t(x) \neq y} 1$
6: **for** $i = 1$ to $(\frac{T}{2} - 1)$ **do**
7:     **if** D="disagreement" OR D="kohavi wolpert variance" OR D="generalized diversity" **then**
8:         $M' = M' \bigcup \underset{M_t \notin M'}{\arg\max} \left[div\left(S, D, M' \cup M_t\right)\right]$
9:     **else**
10:         $M' = M' \bigcup \underset{M_t \notin M'}{\arg\min} \left[div\left(S, D, M' \cup M_t\right)\right]$
11:     **end if**
12: **end for**

Fig. 4.1   Bagging using diversity.

Calculating the Disagreement Measure

Require: S (the original training set), M' (examined set of classifiers)

1: $sum=0$
2: **for** each unique pair of base classifiers $\{a, b\} \subseteq M' | a \neq b$ **do**
3:     $sum=sum +$ number of instances where classifiers a and b disagree
4: **end for**
5: return $\frac{2sum}{|S| \times |M'| \times (|M'|-1)}$

Fig. 4.2   Calculating the disagreement measure.

The successive classifiers are trained on an artificial set that combines tuples from the original training set and also on some fabricated tuples. In each iteration, the input attribute values of the fabricated tuples are generated according to the original data distribution. On the other hand, the target values of these tuples are selected so as to differ maximally from the current ensemble predictions. Comprehensive experiments have demonstrated that this technique is consistently more accurate than the base classifier,

Calculating the Double Fault Measure

Require: S (the original training set), M' (examined set of classifiers)
1: $sum=0$
2: **for** each unique pair of base classifiers $\{a,b\} \subseteq M'|a \neq b$ **do**
3:    $sum=sum$ + number of instances that where misclassified by both a and b.
4: **end for**
5: return $\frac{2sum}{|S|\times|M'|\times(|M'|-1)}$

Fig. 4.3   Calculating the double fault measure.

Calculating the Kohavi Wolpert Variance

Require: S (the original training set), M' (examined set of classifiers)
1: $sum=0$
2: **for** each $\langle x_i, y_i \rangle \in S$ **do**
3:    l_i = number of classifiers misclassified x_i
4:    $sum = sum + l_i \times (|M'| - l_i)$
5: **end for**
6: return $\frac{sum}{|S|\times|M'|^2}$

Fig. 4.4   Calculating the Kohavi Wolpert variance.

Calculating the Inter Rater Measure

Require: S (the original training set), M' (examined set of classifiers)
1: $sum=0$
2: **for** each $\langle x_i, y_i \rangle \in S$ **do**
3:    l_i = number of classifiers misclassified x_i
4:    $sum = sum + l_i \times (|M'| - l_i)$
5: **end for**
6: $P = 1 - \frac{\sum l_i}{|S|\times|M'|}$
7: return $1 - \frac{sum}{|S|\times|M'|\times(|M'|-1)\times P\times(1-P)}$

Fig. 4.5   Calculating the inter rater measure.

Bagging and Random Forests. Decorate also obtains higher accuracy than boosting on small training sets, and achieves comparable performance on larger training sets.

Calculating Generalized Diversity

Require: S (the original training set), M' (examined set of classifiers)
1: **for** each $\langle x_i, y_i \rangle \in S$ **do**
2:     l_i = number of classifiers misclassified x_i
3:     $V_i = \frac{|M'| - l_i}{|M'|}$
4: **end for**
5: return variance of V_i

Fig. 4.6   Calculating generalized diversity.

4.3.3   *Partitioning*

Some argue that classic ensemble techniques (such as boosting and bagging) have limitations on massive datasets, because the size of the dataset can become a bottleneck [Chawla *et al.* (2004)]. Moreover, it is suggested that partitioning the datasets into random, disjoint partitions will not only overcome the issue of exceeding memory size, but will also lead to creating an ensemble of diverse and accurate classifiers, each built from a disjoint partition but with the aggregate processing all of the data. This can improve performance in a way that might not be possible by subsampling. More recently a framework for building thousands of classifiers that are trained from small subsets of data in a distributed environment was proposed [Chawla *et al.* (2004)]. The robust learning from bites (RLB) algorithm that was proposed by Christmann *et al.* [Christmann *et al.* (2007)] is also designed to work with large data sets.

Clustering techniques can be used to partitioning the sample. The goal of clustering is to groups the data instances into subsets in such a manner that similar instances are grouped together, while different instances belong to different groups. The instances are thereby organized into an efficient representation that characterizes the population being sampled. Formally, the clustering structure is represented as a set of subsets $C = C_1, \ldots, C_k$ of S, such that: $S = \bigcup_{i=1}^{k} C_i$ and $C_i \cap C_j = \emptyset$ for $i \neq j$. Consequently, any instance in S belongs to exactly one and only one subset.

The simplest and most commonly used algorithm, employing a squared error criterion is the K-means algorithm. This algorithm partitions the data into K clusters $(C_1, C_2, \ldots, C_K)$, represented by their centers or means. The center of each cluster is calculated as the mean of all the instances belonging to that cluster. The CBCD (cluster-based concurrent decomposition) algorithm [Rokach *et al.*   (2005),] first clusters the instance space by using the K-means clustering algorithm. Then, it creates disjoint sub-samples using the clusters in such a way that each sub-sample is comprised of tuples from all clusters and hence represents the entire dataset. An inducer is applied in turn to each sub-sample. A voting mechanism is used to combine the classifiers classifications. Experimental study indicates that the CBCD algorithm outperforms the bagging algorithm.

Ahn *et al.* [Ahn *et al.* (2007)] indicate that random partition the input attribute set into several subset such that each classifier is induced from a different subset, is particularly useful for high-dimensional datasets. Their experiments indicate that for unbalanced data, this partition approach maintains the balance between sensitivity and specificity more adequately than many other classification methods.

Denison *et al.* [Denison *et al.* (2002)] examine two schemas for partitioning the instance space into disjoint subspaces: The BPM (Bayesian partition model) schema has been shown to be unsuitable when there training set is large or there are many input attributes. The PPM (product partition model) schema provides good results in several cases especially in datasets where there are many irrelevant input attributes and it is less suitable to situations where there are strong interactions among input attributes.

4.4 Manipulating the Target Attribute Representation

In methods that manipulate the target attribute, instead of inducing a single complicated classifier, several classifiers with different and usually simpler representations of the target attribute are induced. This manipulation can be based on an aggregation of the original target's values (known as *Concept Aggregation*) or more complicated functions (known as *Function Decomposition*).

Classical concept aggregation replaces the original target attribute with a function, such that the domain of the new target attribute is smaller than the original one [Buntine (1996)].

The idea to convert K class classification problems into K-two class classification problems has been proposed by [Anand *et al.* (1995)]. Each problem considers the discrimination of one class to the other classes. Lu and Ito [Lu and Ito (1999)] extend Anand's method and propose a new method for manipulating the data based on the class relations among the training data. By using this method, they divide a K class classification problem into a series of $K(K-1)/2$ two-class problems where each problem considers the discrimination of one class to each one of the other classes. The researchers used neural networks to examine this idea.

A general concept aggregation algorithm called *Error-Correcting Output Coding* (ECOC) uses a code matrix to decompose a multi-class problem into multiple binary problems [Dietterich and Bakiri (1995)]. ECOC for multi-class classification hinges on the design of the code matrix. Please refer to Section 6.2 for additional details.

A general-purpose function decomposition approach for machine learning was proposed in [Zupan *et al.* (1998)]. According to this approach, attributes are transformed into new concepts in an iterative manner to create a hierarchy of concepts.

4.4.1 *Label Switching*

Breiman (2000) suggests generating an ensemble by using perturbed versions of the training set where the classes of the training examples are randomly switched.

The classifiers generated by this procedure have statistically uncorrelated errors in the training set. The basic idea of this method is choosing for each run a different set of examples so that their class labels will be changed randomly to a different class label.

In contrast to Breiman (2000), Martinez-Munoz and Suarez (2005) suggest that the probability of switching class label is kept constant (i.e. independent of the original label and class distribution) for every training example. This makes it possible to use larger values of the switching rate in unbalanced datasets. Martinez-Munoz and Suarez (2005) show that high accuracy can be achieved with ensembles generated by class switching provided that fairly large ensembles are generated (around 1000 classifiers). This large ensemble may slow down the whole learning process and in some cases even result in system out of memory.

Figure 4.7 presents the pseudocode of label switching. In each iteration, the class value of some instances are randomly switched. An instance is chosen to be the class label changed with a probability of p where p is an input parameter. If an instance has been chosen to have its class changed, the new class is picked randomly from all the other classes with equal probability.

Label Switching

Require: I (a base inducer), T (number of iterations), S (the original training set), p (the rate of label switching).

1: **for** $t = 1$ to T **do**
2:   $S' =$ a copy of S
3:   **for** each $< x, y >$ in S' **do**
4:     $R = $ A new random number
5:     **if** $R < P$ **then**
6:       Randomly pick a new class label y' form $dom(y)$ that is different form the original class of x
7:       Change the class label of x to be y'
8:     **end if**
9:   **end for**
10:   $M_t = I(S')$
11: **end for**

Fig. 4.7 The label switching algorithm.

4.5 Partitioning the Search Space

The idea is that each member in the ensemble explores a different part of the search space. Thus, the original instance space is divided into several sub-spaces. Each sub-space is considered independently and the total model is a (possibly soft) union of such simpler models.

When using this approach, one should decide if the subspaces will overlap. At one extreme, the original problem is decomposed into several mutually exclusive sub-problems, such that each subproblem is solved using a dedicated classifier. In such cases, the classifiers may have significant variations in their overall performance over different parts of the input space [Tumer and Ghosh (2000)]. At the other extreme, each classifier solves the same original task. In such cases, "If the individual classifiers are then appropriately chosen and trained properly, their performances will be (relatively) comparable in any region of the problem space. [Tumer and Ghosh (2000)]". However, usually the sub-spaces may have soft boundaries, namely sub-spaces are allowed to overlap.

There are two popular approaches for search space manipulations: divide and conquer approaches and feature subset-based ensemble methods.

4.5.1 *Divide and Conquer*

In the neural-networks community, Nowlan and Hinton [Nowlan and Hinton (1991)] examined the mixture of experts (ME) approach, which partitions the instance space into several subspaces and assigns different experts (classifiers) to the different subspaces. The subspaces, in ME, have soft boundaries (i.e., they are allowed to overlap). A gating network then combines the experts' outputs and produces a composite decision. An extension to the basic mixture of experts, known as hierarchical mixtures of experts (HME), has been proposed in [Jordan and Jacobs (1994)]. This extension decomposes the space into sub-spaces, and then recursively decomposes each sub-space into sub-spaces.

Some researchers have used clustering techniques to partition the space [Rokach *et al.* (2003),]. The basic idea is to partition the instance space into mutually exclusive subsets using K-means clustering algorithm. An analysis of the results shows that the proposed method is well suited for datasets of numeric input attributes and that its performance is influenced by the dataset size and its homogeneity.

NBTree [Kohavi (1996)] is an instance space decomposition method that induces a decision tree and a Naïve Bayes hybrid classifier. To induce an NBTree, the instance space is recursively partitioned according to attributes values. The result of the recursive partitioning is a decision tree whose terminal nodes are Naïve Bayes classifiers. Since subjecting a terminal node to a Naïve Bayes classifier means that the hybrid classifier may classify two instances from a single hyper-rectangle region into distinct classes, the NBTree is more flexible than a pure decision tree. More recently Cohen *et al.* (2007) generalizes the NBTree idea and examines a decision-tree framework for space decomposition. According to this framework, the original instance-space is hierarchically partitioned into multiple subspaces and a distinct classifier (such as neural network) is assigned to each subspace. Subsequently, an unlabeled, previously-unseen instance is classified by employing the classifier that was assigned to the subspace to which the instance belongs.

Altincay [Altincay (2007)] propose the use of model ensemble-based nodes where a multitude of models are considered for making decisions at each node. The ensemble members are generated by perturbing the model parameters and input attributes. In generating model ensembles multilayer perceptron (MLP), linear multivariate perceptron and Fishers linear discriminant type models are considered. The first node to be generated is the root node where the data being considered is the whole training set. The algorithm then creates an ensemble of classifiers using random projection. However, in the random subspace approach, the individual accuracies achieved by some of the base classifiers may be insufficient since some of the features may be irrelevant to the learning target. In order to avoid this, the best third classifiers having the highest individual accuracies are selected. The training instances reaching that node are divided into several branches according to the classification provided by the ensemble using majority voting. Then, the algorithm is recursively executed on each branch. The algorithm stops when the number of instances in the current node are below a given threshold. One of the main strengths of the proposed approach is that it uses small number of training samples that reach at nodes close to the leafs in an efficient way. Experiments conducted on several datasets and three model types indicate that the proposed approach achieves better classification accuracies compared to individual nodes, even in cases when only one model class is used in generating ensemble members.

The divide and conquer approach includes many other specific methods such as local linear regression, CART/MARS, adaptive subspace models, etc [Johansen and Foss (1992); Ramamurti and Ghosh (1999)].

4.5.2 *Feature Subset-based Ensemble Methods*

Another less common strategy for manipulating the search space is to manipulate the input attribute set. Feature subset based ensemble methods are those that manipulate the input feature set for creating the ensemble members. The idea is to simply give each classifier a different projection of the training set. Tumer and Oza. [Tumer and Oza (2003)] claim that feature subset-based ensembles potentially facilitate the creation of a classifier for high dimensionality data sets without the feature selection drawbacks mentioned above. Moreover, these methods can be used to improve the classification performance due to the reduced correlation among the classifiers. Bryll et al. [Bryll et al. (2003)] also indicate that the reduced size of the dataset implies faster induction of classifiers. Feature subset avoids the class under-representation which may happen in instance subsets methods such as bagging. There are three popular strategies for creating feature subset-based ensembles: random-based, reduct-based and collective-performance-based strategy.

4.5.2.1   Random-based Strategy

The most straightforward techniques for creating feature subset-based ensemble are based on random selection. Ho [Ho (1998)] uses random subspaces to create forest of decision trees. The ensemble is constructed systematically by pseudo-randomly selecting subsets of features. The training instances are projected to each subset and a decision tree is constructed using the projected training samples. The process is repeated several times to create the forest. The classifications of the individual trees are combined by averaging the conditional probability of each class at the leaves (distribution summation). Ho shows that simple random selection of feature subsets may be an effective technique because the diversity of the ensemble members compensates for their lack of accuracy. Furthermore, random subspace methods are effective when the number of training instances is comparable to number of features.

Bay [Bay (1999)] proposed MFS which uses simple voting in order to combine outputs from multiple KNN (K-Nearest Neighbor) classifiers, each having access only to a random subset of the original features. Each classifier employs the same number of features. This procedure resembles the random subspaces methods.

Bryll *et al.* [Bryll *et al.* (2003)] introduce attribute bagging (AB) which combine random subsets of features. AB first finds an appropriate subset size by a random search in the feature subset dimensionality. It then randomly selects subsets of features, creating projections of the training set on which the classifiers are trained. A technique for building ensembles of simple Bayesian classifiers in random feature subsets was also examined [Tsymbal and Puuronen (2002)] for improving medical applications.

4.5.2.2   Reduct-based Strategy

A reduct is defined as the smallest feature subset which has the same predictive power as the whole feature set. By definition, the size of the ensembles that were created using reducts are limited to the number of features. There have been several attempts to create classifier ensembles by combining several reducts. Wu *et al.* [Wu *et al.* (2005)] introduce the worst-attribute-drop-first algorithm to find a set of significant reducts and then combine them using naïve Bayes. Bao and Ishii [Bao and Ishii (2002)] examine the idea of combining multiple K-nearest neighbor classifiers for text classification by reducts. Hu *et al.* [Hu *et al.* (2005)] propose several techniques to construct decision forests, in which every tree is built on a different reduct. The classifications of the various trees are combined using a voting mechanism.

4.5.2.3 *Collective-Performance-based Strategy*

Cunningham and Carney [Cunningham and Carney (2000)] introduced an ensemble feature selection strategy that randomly constructs the initial ensemble. Then, an iterative refinement is performed based on a hill-climbing search in order to improve the accuracy and diversity of the base classifiers. For all the feature subsets, an attempt is made to switch (include or delete) each feature. If the resulting feature subset produces a better performance on the validation set, that change is kept. This process is continued until no further improvements are obtained. Similarly, Zenobi and Cunningham [Zenobi and Cunningham (2001)] suggest that the search for the different feature subsets will not be solely guided by the associated error but also by the disagreement among the ensemble members.

Tumer and Oza [Tumer and Oza (2003)] present a new method called input decimation (ID), which selects feature subsets based on the correlations between individual features and class labels. This experimental study shows that ID can outperform simple random selection of feature subsets.

Tsymbal *et al.* [Tsymbal *et al.* (2004)] compare several feature selection methods that incorporate diversity as a component of the fitness function in the search for the best collection of feature subsets. This study shows that there are some datasets in which the ensemble feature selection method can be sensitive to the choice of the diversity measure. Moreover, no particular measure is superior in all cases.

Gunter and Bunke [Gunter and Bunke (2004)] suggest employing a feature subset search algorithm in order to find different subsets of the given features. The feature subset search algorithm not only takes the performance of the ensemble into account, but also directly supports diversity of subsets of features.

Combining genetic search with ensemble feature selection was also examined in the literature. Opitz and Shavlik [Opitz and Shavlik (1996)] applied GAs to ensembles using genetic operators that were designed explicitly for hidden nodes in knowledge-based neural networks. In a later research, Opitz [Opitz (1999)] used genetic search for ensemble feature selection. This genetic ensemble feature selection (GEFS) strategy begins by creating an initial population of classifiers where each classifier is generated by randomly selecting a different subset of features. Then, new candidate classifiers are continually produced by using the genetic operators of crossover and mutation on the feature subsets. The final ensemble is composed of the most fitted classifiers. Similarly, the genetic algorithm that Hu *et al.* [Hu *et al.* (2005)] use for selecting the reducts to be included in the final ensemble, first creates N reducts then it trains N decision trees using these reducts. It finally uses a GA for selecting which of the N decision trees are included in the final forest.

4.5.2.4 *Feature Set Partitioning*

Partitioning means dividing the original training set into smaller training sets. A different classifier is trained on each sub-sample. After all classifiers are constructed, the models are combined in some fashion [Maimon and Rokach (2005)]. There are two obvious ways to partition the original dataset: Horizontal Partitioning and Vertical Partitioning. In horizontal partitioning the original dataset is partitioned into several datasets that have the same features as the original dataset, each containing a subset of the instances in the original. In vertical partitioning the original dataset is partitioned into several datasets that have the same number of instances as the original dataset, each containing a subset of the original set of features.

In order to illustrate the idea of partitioning, recall the training set in Table 1.1 which contains a segment of the Iris dataset. This is one of the best known datasets in the pattern recognition literature. The goal in this case is to classify flowers into the Iris subgeni according to their characteristic features. The dataset contains three classes that correspond to three types of iris flowers: $dom(y) = \{IrisSetosa, IrisVersicolor, IrisVirginica\}$. Each pattern is characterized by four numeric features (measured in centimeters): $A = \{sepallength, sepalwidth, petallength, petalwidth\}$. Tables 4.1 and 4.2 respectively illustrate mutually exclusive horizontal and vertical partitions of the Iris dataset. Note that despite the mutually exclusiveness, the class attribute must be included in each vertical partition.

Vertical partitioning (also known as feature set partitioning) is a particular case of feature subset-based ensembles in which the subsets are pairwise disjoint subsets. At the same time, feature set partitioning generalizes the task of feature selection which aims to provide a single representative set of features from which a classifier is constructed. Feature set partitioning, on the other hand, decomposes the original set of features into several subsets and builds a classifier for each subset. Thus, a set of classifiers is trained

Table 4.1   Horizontal Partitioning of the Iris Dataset.

Sepal Length	Sepal Width	Petal Length	Petal Width	Class (Iris Type)
5.1	3.5	1.4	0.2	Iris-setosa
4.9	3.0	1.4	0.2	Iris-setosa
6.0	2.7	5.1	1.6	Iris-versicolor

Sepal Length	Sepal Width	Petal Length	Petal Width	Class (Iris Type)
5.8	2.7	5.1	1.9	Iris-virginica
5.0	3.3	1.4	0.2	Iris-setosa
5.7	2.8	4.5	1.3	Iris-versicolor
5.1	3.8	1.6	0.2	Iris-setosa

Table 4.2   Vertical Partitioning of the Iris Dataset.

Petal Length	Petal Width	Class (Iris Type)
1.4	0.2	Iris-setosa
1.4	0.2	Iris-setosa
5.1	1.6	Iris-versicolor
5.1	1.9	Iris-virginica
1.4	0.2	Iris-setosa
4.5	1.3	Iris-versicolor
1.6	0.2	Iris-setosa

Sepal Length	Sepal Width	Class (Iris Type)
5.1	3.5	Iris-setosa
4.9	3.0	Iris-setosa
6.0	2.7	Iris-versicolor
5.8	2.7	Iris-virginica
5.0	3.3	Iris-setosa
5.7	2.8	Iris-versicolor
5.1	3.8	Iris-setosa

such that each classifier employs a different subset of the original feature set. Subsequently, an unlabelled instance is classified by combining the classifications of all classifiers.

Several researchers have shown that the partitioning methodology can be appropriate for classification tasks with a large number of features [Rokach (2006); Kusiak (2000)]. The search space of a feature subset-based ensemble contains the search space of feature set partitioning, and the latter contains the search space of feature selection.

In the literature there are several works that deal with feature set partitioning. In one research, the features are grouped according to the feature type: nominal value features, numeric value features and text value features [Kusiak (2000)]. A similar approach was also used for developing the linear Bayes classifier [Gama (2000)]. The basic idea consists of aggregating the features into two subsets: the first subset containing only the nominal features and the second only the continuous features.

In another research, the feature set was decomposed according to the target class [Tumer and Ghosh (1996)]. For each class, the features with low correlation relating to that class were removed. This method was applied on a feature set of 25 sonar signals where the target was to identify the meaning of the sound (whale, cracking ice, etc.).

The feature set decomposition can be obtained by grouping features based on pairwise mutual information, with statistically similar features assigned to the same group [Liao and Moody (2000)]. For this purpose one can use an existing hierarchical clustering algorithm. As a consequence, several feature subsets are constructed by selecting one feature from each group. A neural network is subsequently constructed for each subset. All networks are then combined.

In statistics literature, the well-known feature-oriented ensemble algorithm is the MARS algorithm [Friedman (1991)]. In this algorithm, a multiple regression function is approximated using linear splines and their tensor products. It has been shown that the algorithm performs an ANOVA decomposition, namely, the regression function is represented as a grand total of several sums. The first sum is of all basic functions that involve only a single attribute. The second sum is of all basic functions that involve exactly two attributes, representing (if present) two-variable interactions. Similarly, the third sum represents (if present) the contributions from three-variable interactions, and so on. In a recent study, several methods for combining different feature selection results have been proposed [Chizi et al. (2002)]. The experimental results indicate that combining different feature selection methods can significantly improve the accuracy results.

The EROS (Ensemble Rough Subspaces) algorithm is a rough-set-based attribute reduction algorithm [Hu et al. (2007)]. It uses an accuracy-guided forward search strategy to sequentially induce base classifiers. Each base classifier is trained on a different reduct of the original data set. Then a post-pruning strategy is employed to filter out non-useful base classifiers. Experimental results show that EROS outperforms bagging and random subspace methods in terms of accuracy and size of ensemble systems.

A general framework that searches for helpful feature set partitioning structures has also been proposed [Rokach and Maimon (2005b)]. This framework nests many algorithms, two of which are tested empirically over a set of benchmark datasets. This work indicates that feature set decomposition can increase the accuracy of decision trees. More recently, genetic algorithm has been successfully applied for feature set partitioning [Rokach (2008a)]. This GA uses a new encoding schema and a Vapnik-Chervonenkis dimension bound for evaluating the fitness function. The algorithm also suggest a new caching mechanism to speed up the execution and avoid recreation of the same classifiers.

4.5.2.5 *Rotation Forest*

Rotation Forest is an ensemble generation method which aims at building accurate and diverse classifiers [Rodriguez (2006)]. The main idea is to apply feature extraction to subsets of features in order to reconstruct a full feature set for each classifier in the ensemble. Rotation Forest ensembles tend to generate base classifiers which are more accurate than those created by AdaBoost and by Random Forest, and more diverse than those created by bagging. Decision trees were chosen as the base classifiers because of their sensitivity to rotation of the feature axes, while remaining very accurate. Feature extraction is based on principal components analysis (PCA) which is a valuable diversifying heuristic.

Figure 4.8 presents the Rotation Forest pseudocode. For each one of the T base classifiers to be built, we divide the feature set into K disjoint subsets $F_{i,j}$ of equal size M. For every subset, we randomly select a nonempty subset of classes and then draw a bootstrap sample which includes $3/4$ of the original sample. Then we apply PCA using only the features in $F_{i,j}$ and the selected subset of classes. The obtained coefficients of the principal components, $a_{i,1}^1, a_{i,1}^2, \ldots$, are employed to create the sparse "rotation" matrix R_i. Finally we use SR_i from training the base classifier M_i. In order to classify an instance, we calculate the average confidence for each class across all classifiers, and then assign the instance to the class with the largest confidence.

Zhang and Zhang (2008) present the RotBoost algorithm which combines the ideas of Rotation Forest and AdaBoost. RotBoost achieves an even lower prediction error than either one of the two algorithms. RotBoost is presented in Figure 4.9. In each iteration a new rotation matrix is generated and used to create a dataset. The AdaBoost ensemble is induced from this dataset.

Rotation Forest

Require: I (a base inducer), S (the original training set), T (number of iterations), K (number of subsets),

1: **for** $i = 1$ to T **do**
2:     Split the feature set into K subsets: $F_{i,j}$ (for j=1..K)
3:     **for** $j = 1$ to K **do**
4:         Let $S_{i,j}$ be the data set S for the for the features in $F_{i,j}$
5:         Eliminate from $S_{i,j}$ a random subset of classes
6:         Select a bootstrap sample from $S_{i,j}$ of size 75% of the number of objects in $S_{i,j}$. Denote the new set by $S'_{i,j}$
7:         Apply PCA on $S'_{i,j}$ to obtain the coefficients in a matrix $C_{i,j}$
8:     **end for**
9:     Arrange the $C_{i,j}$, for j = 1 to K in a rotation matrix R_i as in the equation:

$$R_i = \begin{bmatrix} a_{i,1}^{(1)}, a_{i,1}^{(2)}, \dots, a_{i,1}^{(M_1)} & [0] & \dots & [0] \\ [0] & a_{i,2}^{(1)}, a_{i,2}^{(2)}, \dots, a_{i,2}^{(M_2)} & \dots & [0] \\ \dots & \dots & \dots & \dots \\ [0] & [0] & \dots & a_{i,k}^{(1)}, a_{i,k}^{(2)}, \dots, a_{i,k}^{(Mk)} \end{bmatrix}$$

10:     Construct R_i^a by rearranging the columns of R_i so as to match the order of features in F
11: **end for**
12: Build classifier M_i using (SR_i^a, X) as the training set

Fig. 4.8   The Rotation Forest.

4.6   Multi-Inducers

In Multi-Inducer strategy, diversity is obtained by using different types of inducers [Michalski and Tecuci (1994)]. Each inducer contains an explicit or implicit bias that leads it to prefer certain generalizations over others. Ideally, this multi-inducer strategy would always perform as well as the best of its ingredients. Even more ambitiously, there is hope that this combination of paradigms might produce synergistic effects, leading to levels of accuracy that neither atomic approach by itself would be able to achieve.

Most research in this area has been concerned with combining empirical approaches with analytical methods (see for instance [Towell and Shavlik (1994)]. Woods *et al.* [Woods *et al.* (1997)] combine four types of base

RotBoost

Require: I (a base inducer), S (the original training set), K (number of attribute subsets), T_1 (number of iterations for Rotation Forest), T_2 (number of iterations for AdaBoost).

1: **for** $s = 1, \cdots T_2$ **do**

2:    Use the steps similar in Rotation Forest to compute the rotation matrix, R_s^a and let $S^a = [XR_s^aY]$ be the training set for classifier C_s.

3:    Initialize the weight distribution over S^a as $D_1(i) = 1/N(i = 1, 2, \cdots N)$.

4:    **for** $t = 1, \cdots T_2$ **do**

5:       According to the distribution D_t, perform N extractions randomly f or S^a with replacement to compose a new set S_t^a.

6:       Apply I to S_t^a to train a classifier C_t^a and then compute the error of $C_t^a as\ \epsilon_t = \Pr_{i \sim D_t}(C_t^a(\mathrm{x}_i) \neq y_i) = \sum_{i=1}^N Ind(C_t^a(\mathrm{x}_i) \neq y_i)D_t(i)$.

7:       **if** $\xi i_t > 0.5$ **then**

8:          set $D_t(i) = 1/N(i = 1, 2, \cdots N)$ and continue with the next loop iteration

9:       **end if**

10:       **if** $\epsilon_t = 0\rangle$ **then**

11:          set $\epsilon_t = 10^{-10}$

12:       **end if**

13:       Choose $\alpha_t = \dfrac{1}{2}\ln(\dfrac{1-\epsilon_t}{\epsilon_t})$

14:       Update the distribution D_t over S^a as $D_{t+1}(i) = \dfrac{D_t(i)}{Z_t} \times$
$\begin{cases} e^{-\alpha_t}, \text{if} C_t^a(\mathrm{x}_i) = y_i \\ e^{\alpha_t})\text{if} C_t^a(\mathrm{x}_i) \neq y_i \end{cases}$ where Z_t is a normalization factor being chosen so that D_{t+1} is a probability distribution over S^a.

15:    **end for**

16: **end for**

Fig. 4.9   The RotBoost algorithm.

inducers (decision trees, neural networks, k-nearest neighbor, and quadratic Bayes). They then estimate local accuracy in the feature space to choose the appropriate classifier for a given new unlabeled instance. Wang *et al.* [Wang *et al.* (2004)] examined the usefulness of adding decision trees to an ensemble of neural networks. The researchers concluded that adding a few decision trees (but not too many) usually improved the performance. Langdon *et al.* [Langdon *et al.* (2002)] proposed using Genetic Programming to find an appropriate rule for combining decision trees with neural networks.

Brodley [Brodley (1995b)] proposed the model class selection (MCS) system. MCS fits different classifiers to different subspaces of the instance space, by employing one of three classification methods (a decision-tree, a discriminant function or an instance-based method). In order to select the classification method, MCS uses the characteristics of the underlined training-set, and a collection of expert rules. Brodley's expert-rules were based on empirical comparisons of the methods' performance (i.e., on prior knowledge).

The NeC4.5 algorithm, which integrates decision tree with neural networks [Zhou and Jiang (2004)], first trains a neural network ensemble. Then, the trained ensemble is employed to generate a new training set by replacing the desired class labels of the original training examples with the output from the trained ensemble. Some extra training examples are also generated from the trained ensemble and added to the new training set. Finally, a C4.5 decision tree is grown from the new training set. Since its learning results are decision trees, the comprehensibility of NeC4.5 is better than that of neural network ensembles.

Using several inducers can solve the dilemma which arises from the "no free lunch" theorem. This theorem implies that a certain inducer will be successful only insofar its bias matches the characteristics of the application domain [Brazdil *et al.* (1994)]. Thus, given a certain application, the practitioner need to decide which inducer should be used. Using the multi-inducer obviate the need to try each one and simplifying the entire process.

4.7   Measuring the Diversity

As stated above, it is usually assumed that increasing diversity may decrease ensemble error [Zenobi and Cunningham (2001)]. For regression problems, *variance* is usually used to measure diversity [Krogh and Vedelsby (1995)]. In such cases it can be easily shown that the ensemble error can be reduced by increasing ensemble diversity while maintaining the average error of a single model.

In classification problems, a more complicated measure is required to evaluate the diversity. There have been several attempts to define diversity measure for classification tasks.

In the neural network literature two measures are presented for examining diversity:

- Classification coverage: An instance is covered by a classifier, if it yields a correct classification.
- Coincident errors: A coincident error amongst the classifiers occurs when more than one member misclassifies a given instance.

Based on these two measures, Sharkey [Sharkey and Sharkey (1997)] defined four diversity levels:

- Level 1 - No coincident errors and the classification function is completely covered by a majority vote of the members.
- Level 2 - Coincident errors may occur, but the classification function is completely covered by a majority vote.
- Level 3 - A majority vote will not always correctly classify a given instance, but at least one ensemble member always correctly classifies it.
- Level 4 - The function is not always covered by the members of the ensemble.

Brown *et al.* [Brown *et al.* (2005)] claim that the above four-level scheme provides no indication of how typical the error behavior described by the assigned diversity level is. This claim, especially, holds when the ensemble exhibits different diversity levels on different subsets of instance space.

There are other more quantitative measures which categorize these measures into two types [Brown *et al.* (2005)]: pairwise and non-pairwise. Pairwise measures calculate the average of a particular distance metric between all possible pairings of members in the ensemble, such as Q-statistic [Brown *et al.* (2005)] or kappa-statistic [Margineantu and Dietterich (1997)]. The non-pairwise measures either use the idea of entropy (such as [Cunningham and Carney (2000)]) or calculate a correlation of each ensemble member with the averaged output. The comparison of several measures of diversity has resulted in the conclusion that most of them are correlated [Kuncheva and Whitaker (2003)].

Kuncheva and Whitaker (2003) divide the diversity measures into two categories: pairwise diversity measures and non-pairwise diversity measures. Here we discuss the pairwise diversity measures. For an ensemble of n classifiers the total pairwise diversity measure is calculated as the mean pairwise measure over all $n \cdot (n-1)/2$ pairs of classifiers:

$F_{Total} = \frac{2}{n(n-1)} \sum\limits_{\forall i \neq j} f_{i,j}$ where $f_{i,j}$ is a similarity or diversity measure of two classifiers outputs i and j. Kuncheva and Whitaker (2003) find the following two diversity pairwise measures useful:

(1) The disagreement measure is defined as the ratio between the number of instances on which one classifier is correct and its counterpart is incorrect to the total number of instances: $Dis_{i,j} = \frac{m_{\bar{i}j}+m_{i\bar{j}}}{m_{\bar{i}j}+m_{i\bar{j}}+m_{ij}+m_{\bar{i}\bar{j}}}$ where m_{ij} specifies the number of instances in which both classifier i and classifier j are correct while $m_{\bar{i}\bar{j}}$ indicates the number of instances that are misclassified by both classifiers. Similarly, $m_{i\bar{j}}$ and $m_{\bar{i}j}$ indicate the number of instances in which one classifier has correctly classified the instances but its counterpart has misclassified these instances.

(2) The double-fault measure is defined as the proportion of the cases that have been misclassified by both classifiers: $DF_{i,j} = \frac{m_{\bar{i}\bar{j}}}{m_{\bar{i}j}+m_{i\bar{j}}+m_{ij}+m_{\bar{i}\bar{j}}}$

Instead of measuring the diversity, we can complementarily use the following pairwise similarity measures:

(1) The Q statistics varies between -1 and 1 and is defined as: $Q_{i,j} = \left(m_{ij} \cdot m_{\bar{i}\bar{j}} - m_{i\bar{j}} \cdot m_{\bar{i}j}\right) / \left(m_{ij} \cdot m_{\bar{i}\bar{j}} + m_{i\bar{j}} \cdot m_{\bar{i}j}\right)$. Positive values indicate that the two classifiers are correlated (namely they tend to correctly classify the same instances). A value close to 0 indicates that the classifiers are independent.

(2) The correlation coefficient – The ρ measure is very similar to the Q measure. It has the same numerator as Q measure. Moreover, it always has the same sign but the value magnitude is never greater than the corresponding Q value: $\rho_{i,j} = \frac{\left(m_{ij} \cdot m_{\bar{i}\bar{j}} - m_{i\bar{j}} \cdot m_{\bar{i}j}\right)}{\sqrt{\left(m_{ij}+m_{i\bar{j}}\right) \cdot \left(m_{ij}+m_{\bar{i}j}\right) \cdot \left(m_{\bar{i}\bar{j}}+m_{i\bar{j}}\right) \cdot \left(m_{\bar{i}\bar{j}}+m_{\bar{i}j}\right)}}$

Kuncheva and Whitaker (2003) show that these measures are strongly correlated between themselves. Still on specific real classification tasks, the measures might behave differently, so they can be used as a complementary set. Nevertheless, Kuncheva and Whitaker (2003) could not find a definitive connection between the measures and the improvement of the accuracy. Thus, they conclude that it is unclear if diversity measures have any practical value in building classifier ensembles.

Tang *et al.* (2006) explain the relation between diversity measures and the concept of margin, which is more explicitly related to the success of ensemble learning algorithms. They present the uniformity condition for maximizing both the diversity and the minimum margin of an ensemble and demonstrated theoretically and experimentally the ineffectiveness of the diversity measures for constructing ensembles with good generalization performance. Tang *et al.* (2006) specify three reasons for that:

(1) The alteration in the diversity measures does not afford consistent guidance on whether a set of base classifiers provide low generalization error.
(2) The existing diversity measures are correlated to the mean accuracy of the base classifiers. Thus, they do not provide any additional information to the accuracy measure.
(3) Most of the diversity measures has no regularization term. Thus, even if we maximize their values, we may over-fit the ensemble.

Chapter 5

Ensemble Selection

5.1 Ensemble Selection

An important aspect of ensemble methods is to define how many base classifiers should be used.

Ensemble selection, also known as ensemble pruning or shrinkage aims at dilute the ensemble. There are two main reasons for reducing the ensemble size: a) Reducing computational overhead: Smaller ensembles require less computational overhead and b) Improving Accuracy: Some members in the ensemble may reduce the predictive performance of the whole. Pruning these members can increase the accuracy. Still, in some cases shrinkage can actually cause the ensemble to overfit in a situation where it otherwise would not have [Mease and Wyner (2008)].

There are several factors that may determine this size:

- Desired accuracy — In most cases, ensembles containing ten classifiers are sufficient for reducing the error rate. Nevertheless, there is empirical evidence indicating that: when AdaBoost uses decision trees, error reduction is observed in even relatively large ensembles containing 25 classifiers [Opitz and Maclin (1999)]. In disjoint partitioning approaches, there may be a trade-off between the number of subsets and the final accuracy. The size of each subset cannot be too small because sufficient data must be available for each learning process to produce an effective classifier.
- Computational cost — Increasing the number of classifiers usually increases computational cost and decreases their comprehensibility. For that reason, users may set their preferences by predefining the ensemble size limit.

- The nature of the classification problem - In some ensemble methods, the nature of the classification problem that is to be solved, determines the number of classifiers.
- Number of processors available — In independent methods, the number of processors available for parallel learning could be put as an upper bound on the number of classifiers that are treated in paralleled process.

There are three approaches for determining the ensemble size, as described by the following subsections.

5.2   Pre Selection of the Ensemble Size

This is the most simple way to determine the ensemble size. Many ensemble algorithms have a controlling parameter such as "number of iterations", which can be set by the user. Algorithms such as Bagging belong to this category. In other cases the nature of the classification problem determine the number of members (such as in the case of ECOC).

5.3   Selection of the Ensemble Size While Training

There are ensemble algorithms that try to determine the best ensemble size while training. Usually as new classifiers are added to the ensemble these algorithms check if the contribution of the last classifier to the ensemble performance is still significant. If it is not, the ensemble algorithm stops. Usually these algorithms also have a controlling parameter which bounds the maximum size of the ensemble.

Random forests algorithm uses out-of-bag (oob) procedure to get an unbiased estimate of the test set error [Breiman (1999)]. The effectiveness of using out-of-bag error estimate, to decide when a sufficient number of classification trees have been recently examined in [Banfield et al. (2007)]. Specifically, the algorithm works by first smoothing the out-of-bag error graph with a sliding window in order to reduce the variance. After the smoothing has been completed, the algorithm takes a larger window on the smoothed data points and determines the maximum accuracy within that window. It continues to process windows until the maximum accuracy within a particular window no longer increases. At this point, the stopping criterion has been reached and the algorithm returns the ensemble with the maximum raw accuracy from within that window. It has been shown that out-of-bag obtain an accurate ensemble for those methods that incorporate bagging into the construction of the ensemble.

Mease and Wyner (2008) indicate that using stopping rules, may be harmful in certain cases since they would stop the algorithm after only a few iterations when the overfitting first takes place, despite the fact that the best performance is again achieved after adding more base classifiers.

5.4 Pruning - Post Selection of the Ensemble Size

As in decision tree induction, it is sometimes useful to let the ensemble grow freely and then prune the ensemble in order to get more effective and compact ensembles. Post selection of the ensemble size allows ensemble optimization for such performance metrics as accuracy, cross entropy, mean precision, or the ROC area. Empirical examinations indicate that pruned ensembles may obtain a similar accuracy performance as the original ensemble [Margineantu and Dietterich (1997)]. In another empirical study that was conducted in order to understand the affect of ensemble sizes on ensemble accuracy and diversity, it has been shown that it is feasible to keep a small ensemble while maintaining accuracy and diversity similar to those of a full ensemble [Liu et al., 2004].

The problem of ensemble pruning is to find the best subset such that the combination of the selected classifiers will have the highest possible degree of accuracy. Consequently the problem can be formally phrased as follows:

Given an ensemble $\Omega = \{M_1, \ldots, M_n\}$, a combination method C, and a training set S from a distribution D over the labeled instance space, the goal is to find an optimal subset $Z_{opt} \subseteq \Omega$. which minimizes the generalization error over the distribution D of the classification of classifiers in Z_{opt} combined using method C.

Note that we assume that the ensemble is given, thus we do not attempt to improve the creation of the original ensemble.

It has been shown that the pruning effect is more noticeable on ensemble whose the diversity among its members is high (Margineantu and Dietterich, 1997). Boosting algorithms create diverse classifiers by using widely different parts of the training set at each iteration (Zhang *et al.*, 2006). Specifically we employ the most popular methods for creating the ensemble: Bagging and AdaBoost. Bagging (Breiman, 1996) employs bootstrap sampling to generate several training sets and then trains a classifier from each generated training set. Note that, since sampling with replacement is used, some of the original instances may appear more than once in the same generated training set and some may not be included at all. The classifier predictions are often combined via majority voting. AdaBoost (Freund and Schapire, 1996) sequentially constructs a series of classifiers, where the training instances that are wrongly classified by a certain classifier will get a higher weight in the training of its subsequent classifier. The classifiers' predictions are combined via weighted voting where the weights are determined by the algorithm itself based on the training error of each classifier. Specifically the weight of classifier i is determined by Equation 5.1:

$$\alpha_i = \frac{1}{2} \ln \left(\frac{1 - \varepsilon_i}{\varepsilon_i} \right) \tag{5.1}$$

where ε_i is the training error of classifier i.

The ensemble pruning problem resemble to the well known feature selection problem. However, instead of selecting features one should select the ensemble's members (Liu *et al.*, 2004). This lead to the idea of adapting the Correlation-based Feature Selection method (Hall, 2000) to the current problem. The CFS algorithm is suitable to this case, because in many ensembles there are many correlated base-classifiers.

In earlier research on ensemble Pruning (Margineantu and Dietterich, 1997), the goal was to use a small size of ensemble to achieve an equivalent performance of a boosted ensemble. This has been proved to be NP-hard and is even hard to approximate (Tamon and Xiang, 2000), and the pruning may sacrifice the generalization ability of the final ensemble. After Zhou *et al.* (2002) proved the "many-could-be-better-than-all" theorem, it becomes well-known that it is possible to get a small yet strong ensemble. This arose many new ensemble pruning methods. Tsoumakas *et al.* (2008) propose the organization of the various ensemble selection methods into the following categories: a) Search-based, b) Clustering based c) Ranking-based and d) Other.

5.4.1   *Ranking-based*

The idea of this approach is to once rank the individual members according to a certain criterion and choosing the top ranked classifiers according to a threshold. For example Prodromidis *et al.* (1999) suggest ranking classifiers according to their classification performance on a separate validation set and their ability to correctly classify specific classes. Similarly Caruana *et al.* (2004) presented a forward stepwise selection procedure in order to select the most relevant classifiers (that maximize the ensemble's performance) among thousands of classifiers. The algorithm FS-PP-EROS generates a selective ensemble of rough subspaces (Hu *et al.*, 2007). The algorithm performs an accuracy-guided forward search to select the most relevant members. The experimental results show that FS-PP-EROS outperforms bagging and random subspace methods in terms of accuracy and size of ensemble systems. In attribute bagging (Bryll *et al.*, 2003), classification accuracy of randomly selected m-attribute subsets is evaluated by using the wrapper approach and only the classifiers constructed on the highest ranking subsets participate in the ensemble voting. Margineantu and Dietterich (1997) present an agreement based ensemble pruning which measures the Kappa statistics between any pair of classifiers. Then pairs of classifiers are selected in ascending order of their agreement level till the desired ensemble size is reached.

5.4.2 *Search-based Methods*

Instead of separately ranking the members, one can perform a heuristic search in the space of the possible different ensemble subsets while evaluating the collective merit of a candidate subset. The GASEN algorithm was developed for selecting the most appropriate classifiers in a given ensemble (Zhou *et al.*, 2002). In the initialization phase, GASEN assigns a random weight to each of the classifiers. Consequently, it uses genetic algorithms to evolve those weights so that they can characterize to some extent the fitness of the classifiers in joining the ensemble. Finally, it removes from the ensemble those classifiers whose weight is less than a predefined threshold value. A revised version of the GASEN algorithm called GASEN-b has been suggested (Zhou and Tang, 2003). In this algorithm, instead of assigning a weight to each classifier, a bit is assigned to each classifier indicating whether it will be used in the final ensemble. In an experimental study the researchers showed that ensembles generated by a selective ensemble algorithm, which selects some of the trained C4.5 decision trees to make up an ensemble, may be not only smaller in size but also stronger in the generalization than ensembles generated by non-selective algorithms. A similar approach can also be found in (Kim *et al.*, 2002). Rokach *et al.* (2006) suggest first to rank the classifiers according to their ROC performance. Then, they suggest evaluating the performance of the ensemble subset by using the top ranked members. The subset size is increased gradually until there are several sequential points with no performance improvement. Prodromidis and Stolfo (2001) introduce a backwards correlation based pruning. The main idea is to remove the members that are least correlated to a meta-classifier which is trained based on the classifiers outputs. In each iteration they remove one member and recompute the new reduced meta-classifier (with the remaining members). The meta-classifier in this case is used to evaluate the collective merit of the ensemble. Windeatt and Ardeshir (2001) compared several subset evaluation methods that were applied to Boosting and Bagging. Specifically the following pruning methods have been compared: Minimum Error Pruning (MEP), Error-based Pruning (EBP), Reduced-Error Pruning(REP), Critical Value Pruning (CVP) and Cost-Complexity Pruning (CCP). The results indicate that if a single pruning method needs to be selected then overall the popular EBP makes a good choice. Zhang el al. (2006) formulate the ensemble pruning problem as a quadratic integer programming problem to look for a subset of classifiers that has the optimal accuracy-diversity trade-off. Using a semidefinite programming (SDP) technique, they efficiently approximate the optimal solution, despite the fact that the quadratic problem is NP-hard.

Which approach to use? Search Based Methods provide a better classification performance than the ranking based methods (Prodromidis *et al.*, 1999). However Search Based methods are usually computational expensive due to their need for searching a large space. Thus one should select a feasible search strategy. Moreover independently to the chosen search strategy, the computational complexity for a evaluating a single candidate subset usually is at least linear in the number of instances in the training set (see Tsoumakas *et al.*, 2008 for complexity analysis of existing evolution measures.)

5.4.2.1 *Collective Agreement-based Ensemble Pruning Method*

The Collective Agreement-based Ensemble Pruning (CAP) calculates the member-class and member-member agreements based on the training data. Member-class agreement indicates how much the member's classifications agree with the real label while member-member agreement is the agreement between the classifications of two members. The merit of an ensemble subset Z with n_z members can be estimated from:

$$Merit_z = \frac{n_z \overline{\kappa}_{cm}}{\sqrt{n_z + n_z(n_z - 1)\overline{\kappa}_{mm}}} \tag{5.2}$$

where $\overline{\kappa}_{cf}$ is the mean agreement between the Z's members and the class and $\overline{\kappa}_{mm}$ is the average member-member agreements in Z.

Eq. 2 is adopted from test theory. It is mathematically derived (Gulliksen, 1950, pages 74-89) from Spearman formula (Spearman, 1913) for the sake of measuring the augmented validity coefficient of a psychological test which consists of nz unit tests. Later, it has been used in human relations studies for evaluating the validity of aggregating experts' opinions (Hogarth, 1977) which is similar to the problem addressed in this paper. According to Eq. 2 the following properties can be observed:

(1) The lower the inter-correlations among classifiers, the higher the merit value.
(2) The higher the correlations between the classifiers and the real class, the merit value increases.
(3) As the number of classifiers in the ensemble increases (assuming the additional classifiers are the same as the original classifiers in terms of their average intercorrelation with the other classifiers and with the real class), the higher the merit value. However, it is unlikely that a large ensemble of classifiers that are all highly correlated with the real class will at the same time bear low correlations with each other (Hogarth, 1977).

The above properties indicate that removing a classifier from an ensemble might be beneficial. Moreover it might not necessarily pay to remove the classifier with the lowest individual agreement with the real class (if it has low intercorrelation with other members).

Several measures can be incorporated in Eq 2, for measuring the agreement. Specifically, the Kappa statistics can be used to measure the agreement in Eq. 2:

$$\kappa_{i,j} = \frac{\vartheta_{i,j} - \theta_{i,j}}{1 - \theta_{i,j}} \tag{5.3}$$

where $\vartheta_{i,j}$ is the proportion of instances on which the classifiers i and j agree with each other on the training set, and $\theta_{i,j}$ is the probability that the two classifiers agree by chance.

Alternatively one can use the symmetrical uncertainty (a modified information gain measure) to measure the agreement between two members (Hall, 2000):

$$SU_{i,j} = \frac{H(\hat{y}_i) + H(\hat{y}_j) - H(\hat{y}_i, \hat{y}_j)}{H(\hat{y}_i) + H(\hat{y}_j)} \tag{5.4}$$

where $\hat{y}_i$ is the classification vector of classifier i and H is the entropy function.

Both Kappa statistics and the Entropy measure have been previously mentioned in the ensemble literature (Kuncheva, 2004). However, they have been merely used for measuring the diversity in classifier ensembles. The novelty of this research relies on the way they are incorporated in the ensemble's merit estimation (Eq 2). Instead of just averaging the agreement measure across all pairs of classifiers ($\overline{\kappa}_{mm}$) and obtain a global pairwise agreement measure. We suggest also taking into consideration the agreement between the classifier's outputs and the real class ($\overline{\kappa}_{cm}$). Thus, the proposed method prefers sub-ensemble whose members have greater agreement with the real class (i.e. more accurate) and have lesser agreement among themselves.

In this sense, the proposed merit measure reminds Breiman's upper bound on the generalization error of random forest (Breiman, 2001) which is expressed *"in terms of two parameters that are measures of how accurate the individual classifiers are and of the agreement between them."* While Brieman's bound is theoretically justified, it is not considered to be very tight (Kuncheva, 2004). Moreover it is merely designed for decision forests.

As the search space is huge (2^n), we are using best first search strategy as the preferred strategy. It explores the search space by making local changes to the current ensemble subset. Best first search strategy begins with an empty ensemble subset. If the path being explored does not achieve an improved merit, the best first strategy backtracks to a more promising previous subset and continues the search from there. The search stops if five consecutive iterations obtain non-improving subsets.

The pseudocode of the proposed algorithm is presented in Figure 5.1. The algorithm gets as input the training set, the ensemble of classifiers, the method for calculating the agreement measure (for example Kappa statistics) and the search strategy. It first calculates the classifiers' output (prediction) on each instance in the training set (Lines 1-5). Then it calculates the mutual agreement matrix among the classifiers' outputs and the agreement between each classifier's output and the actual class (Lines 6-11). Finally it searches the space according to the given search strategy. The search procedure uses the merit calculation for evaluating a certain solution (Lines 14-24).

CAP (S,Q,CT,k)

Input: S Training set

Ω Ensemble of classifiers $\{M_1, \ldots, M_n\}$

Agr A method for calcuating the agreement measure

Src A search startegy

Output: Z Pruned ensemble set

1: FOR each $< x_q, y_q > \in S$ /* Getting members' classifications */
2: FOR each $M_i \in \Omega$

$$3 : \hat{y}_{i,q} \leftarrow M_i(x_q)$$

4: END FOR
5: END FOR
6: FOR each $M_i \in \Omega$ /* Preparing the agreement matrix */

$$7 : CM_i = Agr\,(y, \hat{y}_i)$$

8: FOR each $M_j \in \Omega\,;\, j > i$

$$9 : MM_{i,j} = Agr\,(\hat{y}_i, \hat{y}_j)$$

10: END FOR
11: END FOR
12: $Z \leftarrow Src(\Omega, MM, CM)$ /* Searching the space using the merit function */
13: Return Z

EvaluateMerit (Z,CM,MM)

Input: Z The Ensemble Subset

CM Class-Member agreement vecor

MM Member-Member agreement matrix

Output: $Merit_z$ - The merit of Z.

$$14 : n_z \leftarrow |Z|$$

$$15 : \overline{\kappa}_{cm} \leftarrow 0$$

$$16 : \overline{\kappa}_{mm} \leftarrow 0$$

17: FOR each $M_i \in Z$

$$18 : \overline{\kappa}_{cm} \leftarrow \overline{\kappa}_{cm} + CM_i$$

19: FOR each $M_j \in Z; j > i$

$$20 : \overline{\kappa}_{mm} \leftarrow \overline{\kappa}_{mm} + MM_{i,j}$$

21: END FOR
22: END FOR

$$23 : Merit_z \leftarrow \frac{n_z \overline{\kappa}_{cm}}{\sqrt{n_z + n_z(n_z - 1)\overline{\kappa}_{mm}}}$$

24: Return $Merit_z$

Fig. 5.1   A pseudocode of collective agreement-based pruning of ensembles.

The computational complexity of the agreement matrix calculation (lines 6-11) is $o(n^2 m)$ assuming that the complexity of the agreement measure is $o(m)$. This assumption is true for the two measures presented in equations 3 and 4. The computational complexity of the merit evaluation (lines 14-24) is: $o(n^2)$. If the search strategy imposes a partial ordering on the search space, then the merit can be calculated incrementally. For example if backward search is used then it is requires one addition to the numerator and up to n additions/subtractions in the denominator.

Note that the actual computational complexity depends on the computational complexity of the classifier making a classification (line 3) and the computational complexity of the search strategy which is being used (line 12). Nevertheless neither the computational complexity of evaluating a solution's merit nor the search space size depends on the training set size. Thus, the proposed method makes it possible to thoroughly search the space for problems with large training sets. For example the complexity for a forward selection or backward elimination is $o(n^2)$. Best first search is exhaustive, but the use of a stopping criterion makes the probability of exploring the entire search space small.

5.4.3 Clustering-based Methods

Clustering-based methods have two stages. In the first phase, a clustering algorithm is used in order to discover groups of classifiers that make similar classifications. Then in the second phase, each group of classifiers is separately pruned in order to ensure the overall diversity of the ensemble.

Lazarevic and Obradovic (2001) use the well-known k-means algorithm to perform the clustering of classifiers. They iteratively increase k until the diversity between them starts to decrease. In the second phase they prune the classifiers of each cluster by considering the classifiers in turn from the least accurate to the most accurate. A classifier is kept in the ensemble if its disagreement with the most accurate classifier is more than a predefined threshold and is sufficiently accurate.

Giacinto *et al.* (2000) use Hierarchical Agglomerative Clustering (HAC) for identifying the groups of classifiers. HAC returns a hierarchy of different clustering results starting from as many clusters as classifiers and ending at a single cluster which is identical to the original ensemble. In order to create the hierarchy, they define a distance metric between two classifiers as the probability that the classifiers do not make coincident misclassification and estimate it from a validation set. A distance between two groups is defined as the maximum distance between two classifiers belonging to these clusters. In the second phase, they prune each cluster by selecting the single best performing classifier. Similarly Fu *et al.* (2005) use k-means algorithm to cluster the classifiers into groups but then select the best classifier from each cluster.

5.4.4 Pruning Timing

The pruning methods can be divided into two groups: pre-combining pruning methods and post-combining pruning methods.

5.4.4.1 Pre-combining Pruning

Pre-combining pruning is performed before combining the classifiers. Classifiers that seem to perform well are included in the ensemble. Prodromidis *et al.* [Prodromidis *et al.* (1999)] present three methods for pre-combining pruning: based on an individual classification performance on a separate validation set, diversity metrics, the ability of classifiers to classify correctly specific classes.

In attribute bagging [Bryll *et al.* (2003)], classification accuracy of randomly selected m-attribute subsets is evaluated by using the wrapper approach and only the classifiers constructed on the highest ranking subsets participate in the ensemble voting.

5.4.4.2 *Post-combining Pruning*

In post-combining pruning methods, we remove classifiers based on their contribution to the collective.

Prodromidis [Prodromidis *et al.* (1999)] examines two methods for post-combining pruning assuming that the classifiers are combined using meta-combination method: Based on decision tree pruning and the correlation of the base classifier to the unpruned meta-classifier.

A forward stepwise selection procedure can be used in order to select the most relevant classifiers (that maximize the ensemble's performance) among thousands of classifiers [Caruana *et al.* (2004)]. It has been shown that for this purpose one can use feature selection algorithms. However, instead of selecting features one should select the ensemble's members [Liu et al., 2004].

Rokach *et al.* [Rokach *et al.* (2006),] suggest first to rank the classifiers according to their ROC performance. Then, they suggest to plot a graph where the Y- axis displays a performance measure of the integrated classification . The X-axis presents the number of classifiers that participated in the combination. i.e., the first best classifiers from the list are combined by voting (assuming equal weights for now) with the rest getting zero weights. The ensemble size is chosen when there are several sequential points with no improvement.

The algorithm FS-PP-EROS generates a selective ensemble of rough subspaces [Hu *et al.* (2007)]. The algorithm performs an accuracy-guided forward search and post-pruning strategy to select part of the base classifiers for constructing an efficient and effective ensemble system. The experimental results show that FS-PP-EROS outperform bagging and random subspace methods in terms of accuracy and size of ensemble systems.

The GASEN algorithm was developed for selecting the most appropriate classifiers in a given ensemble [Zhou *et al.* (2002)]. In the initialization phase, GASEN assigns a random weight to each of the classifiers. Consequently, it uses genetic algorithms to evolve those weights so that they can characterize to some extent the fitness of the classifiers in joining the ensemble. Finally, it removes from the ensemble those classifiers whose weight is less than a predefined threshold value.

Recently a revised version of the GASEN algorithm called GASEN-b has been suggested [Zhou and Tang (2003)]. In this algorithm, instead of assigning a weight to each classifier, a bit is assigned to each classifier indicating whether it will be used in the final ensemble. In an experimental study the researchers showed that ensembles generated by a selective ensemble algorithm, which selects some of the trained C4.5 decision trees to make up an ensemble, may be not only smaller in size but also stronger in the generalization than ensembles generated by non-selective algorithms.

A study had compared several post combining pruning methods that were applied to Boosting and Bagging [Windeatt and Ardeshir (2001)]. Specifically the following pruning methods have been compared: Minimum Error Pruning (MEP), Error-based Pruning (EBP), Reduced-Error Pruning(REP), Critical Value Pruning (CVP) and Cost-Complexity Pruning (CCP). The results indicate that if a single pruning method needs to be selected then overall the popular EBP makes a good choice.

A comparative study of pre combining pruning and post combining pruning methods when meta-combining methods are used has been performed in [Prodromidis *et al.* (1999)]. The results indicate that the post-combining pruning methods tend to perform better in this case.

Zhang *et al.* [Zhang *et al.* (2009)] use boosting for determining the order in which the base classifiers are fused, and then construct a pruned ensemble by stopping the fusion process early. Two heuristics rules are used to stop fusion: one is to select the upper twenty percent of the base classifiers from the ordered full Double-Bagging ensemble and the other is to stop the fusion when the weighted training error reaches 0.5.

Croux *et al.* [Croux *et al.* (2007)] propose the idea of trimmed bagging which aims to prune classifiers that yield the highest error rates, as estimated by the out-of-bag error rate. It has been shown that trimmed bagging performs comparably to standard bagging when applied to unstable classifiers as decision trees, but yields improved accuracy when applied to more stable base classifiers, like support vector machines.

Chapter 6

Error Correcting Output Codes

Some machine learning algorithms are designed to solve binary classification tasks, i.e. to classify an instance into only two classes. For example in the direct marketing scenario, a binary classifier can be used to classify potential customers as to whether they will positively or negatively respond to a particular marketing offer.

In many real problems, however, we are required to differentiate between more than two classes. Examples of such problems are the classification of handwritten letters [Knerr *et al.* (1992)], differentiating between multiple types of cancer [Statnikov *et al.* (2005)] and text categorization [Berger (1999); Ghani (2000)].

A multiclass classification task is essentially more challenging than a binary classification task, since the induced classifier must classify the instances into a larger number of classes, which also increases the likelihood for misclassification. Let us consider, for example, a balanced classification problem, with a similar number of data per class, with equiprobable classes and a random classifier. If the problem is binary, the probability of obtaining a correct classification is 50%. For four classes, this probability drops to 25%.

Several machine learning algorithms, such as SVM [Cristianini and Shawe-Taylor (2000)], were originally designed to solve only binary classification tasks. There are two main approaches for applying such algorithms to multiclass tasks. The first approach, which involves extending the algorithm, has been applied to SVMs [Weston and Watkins (1999)] or boosting[Freund and Schapire (1997)]. However, extending these algorithms into a multiclass version may be either impractical or, frequently, not easy to perform [Passerini *et al.* (2004)]. For SVMs, in particular, Hsu and Lin (2002) observed that the reformulation of this technique into multiclass versions leads to high costs in training algorithms.

The second approach is to convert the multiclass task into an ensemble of binary classification tasks, whose results are then combined. The decomposition that has been performed can be generally represented by a code-matrix $\vec{M}$ [Allwein *et al.* (2000)]. There are several alternatives to decomposing the multiclass problem into binary subtasks [Allwein *et al.* (2000)].This matrix has k rows, representing codewords ascribed to each of the k classes in the multiclass task; the columns correspond to the desired outputs of the binary classifiers induced in the decomposition.

In order to illustrate the idea of multiclass decomposition, recall the training set in Table 1.1 which contains a segment of the Iris dataset, one of the best known datasets in pattern recognition literature. The goal in this case is to classify flowers into Iris subspecies according to their characteristic features. The dataset contains three classes that correspond to three types of irises: $dom\,(y) = \{IrisSetosa, IrisVersicolor, IrisVirginica\}$. Table 6.1 illustrates a code-matrix for the Iris dataset. It contains one row for each class and one column for each classifier to be built. The first classifier attempts to distinguish between $\{IrisSetosa, IrisVirginica\}$, which is represented by the binary value of 1 in column 1 and $\{IrisVersicolor\}$, represented by the value -1 in column 1. Similarly, the second classifier attempts to distinguish between $\{IrisVersicolor, IrisVirginica\}$, represented by the binary value of 1 in column 2 and $\{IrisSetosa\}$, represented by the value -1 in column 2. Finally, the third classifier attempts to distinguish between $\{IrisSetosa, IrisVersicolor\}$, represented by the binary value of 1 in column 3 and $\{IrisVirginica\}$, represented by the value -1 in column 3.

Table 6.1   Illustration of Code-matrix for the Iris dataset.

Class Label	First Classifier	Second Classifier	Third Classifier
Iris Setosa	1	-1	1
Iris Versicolor	-1	1	1
Iris Virginica	1	1	-1

In order to classify a new instance into one of the three classes, we first obtain the binary classification from each of the base classifiers. Based on these binary classifications, we search for the most probable class. We simply measure the Hamming distance of the obtained code to the codewords ascribed to each class. The class with the shortest distance is chosen to be the output class. In case of a tie, we arbitrarily select the class. This process is also known as decoding. Table 6.2 presents the Hamming distance for each possible output of an unseen instance. For example, if a certain instance is classified as $-1, 1, -1$ by classifiers 1 to 3 respectively, then its predicted class is either Versicolor or Virginica.

Table 6.2   Decoding Process for the Iris dataset.

Output			Hamming Distance			
Classifier 1	Classifier 2	Classifier 3	Setosa	Versicolor	Virginica	Predicted Class
-1	-1	-1	4	4	4	Any Class
-1	-1	1	2	2	6	Setosa OR Versicolor
-1	1	-1	6	2	2	Versicolor OR Virginica
-1	1	1	4	0	4	Versicolor
1	-1	-1	2	6	2	Setosa OR Virginica
1	-1	1	0	4	4	Setosa
1	1	-1	4	4	0	Virginica
1	1	1	2	2	2	Any Class

6.1   Code-matrix Decomposition of Multiclass Problems

There are several reasons for using decomposition tactics in multiclass solutions. Mayoraz and Moreira (1996), Masulli and Valentini (2000) and Frnkranz (2002) state that implementing a decomposition approach may lessen the computational complexity required for inducing the classifier. Thus, even multiclass induction algorithms may benefit from converting the problem into a set of binary tasks. Knerr *et al.* (1992), claim that the classes in a digit recognition problem (such as the LED problem) could be linearly separated when the classes are analyzed in pairs. Consequently , they combine linear classifiers for all pairs of classes. This is simpler than using a single multiclass classifier that separates all classes simultaneously. Pimenta and Gama (2005) claims that the decomposition approach suggests new possibilities for parallel processing, since the binary subproblems are independent and can be solved in different processors.

Crammer and Singer (2002) differentiate between three sub-categories:

(1) Type I - Given a code matrix, find a set of binary classifiers that result in a small empirical loss;
(2) Type II - Find simultaneously both a set of binary classifiers and a code matrix that produces a small empirical loss.
(3) Type III – Given a set of binary classifiers, find a code matrix that yields a small empirical loss;

In this chapter we focus on type I and II tasks.

6.2   Type I - Training an Ensemble Given a Code-Matrix

The most simple decomposition tactic is the one-against-one (1A1) decomposition also known as the round robin classification. It consists of building $k(k-1)/2$ classifiers, each distinguishing a pair of classes i and j, where $i \neq j$ [Knerr *et al.* (2000); Hastie and Tibshirani (1998)]. To combine the outputs produced by these classifiers, a majority voting scheme can be applied [Kreβel (1999)]. Each 1A1 classifier provides one vote to its preferred class. The final result is the class with most of the votes. Table 6.3 illustrates the 1AA matrix for a four-class classification problem.

In certain cases, only a subset of all possible pairwise classifiers should be used [Cutzu (2003)]. In such cases it is preferable to count the "against" votes, instead of counting the "for" votes since there is a greater chance of making a mistake when counting the latter. If a certain classifier attempts to differentiate between two classes, neither of which is the true class, counting the "for" votes inevitably causes a misclassification. On the other hand, voting against one of the classes will not result in misclassification in such cases. Moreover, in the 1A1 approach, the classification of a classifier for a pair of classes (i, j) does not provide useful information when the instance does not belong to classes i or j [Alpaydin and Mayoraz (1999)]

1A1 decomposition is illustrated in Figure 6.1. For the four-class problem presented in Figure 6.1(a), the 1A1 procedure induces six classifiers, one for each pair of classes. Figure 6.1(b) shows the classifier of class 1 against 4.

Table 6.3   Illustration of One-Against-One (1A1) Decomposition for Four Classes Classification Problem.

Class Label	Classifier 1	Classifier 2	Classifier 3	Classifier 4	Classifier 5	Classifier 6
Class 1	1	1	1	0	0	0
Class 2	-1	0	0	1	1	0
Class 3	0	-1	0	-1	0	1
Class 4	0	0	-1	0	-1	-1

Another standard methodology is called one-against-all (1AA). Given a problem with k classes, k binary classifiers are generated. Each binary classifier is responsible for differentiating a class i from the remaining classes. The classification is usually chosen according to the class with the highest probability. Table 6.4 illustrates the 1AA matrix for a four-class classification problem.

Table 6.4 Illustration of One-Against-All Code-matrix for Four
Classes Classification Problem.

Class Label	Classifier 1	Classifier 2	Classifier 3	Classifier 4
Class 1	1	-1	-1	-1
Class 2	-1	1	-1	-1
Class 3	-1	-1	1	-1
Class 4	-1	-1	-1	1

Some difficulties arise in 1AA decomposition when the training set is unbalanced. The number of instances that are associated with a certain class is much smaller than the number of instances in other classes. In such cases it will be difficult to generate a classifier with good predictive performance in the considered class.

Figure 6.1(c) illustrates one of the classifiers using the 1AA procedure, namely, the one that examine class 1 against all other classes. It is obvious that in the 1A1 procedure, each base classifier uses fewer instances and thus according to [Fürnkranz (2002)] "has more freedom for fitting a decision boundary between the two classes".

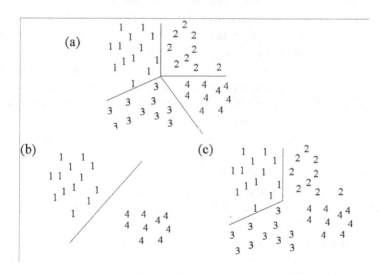

Fig. 6.1 Illustration of 1A1 and 1AA for a four classes problem.

6.2.1   *Error correcting output codes*

Error-correcting systems, which date back to the mid-20th century, have been used to increase the reliability of communication channels. Communication channels suffer from undesired noise that distorts the received message from the original. Due to noise, an arbitrary bit in a message will be changed from its original state. To reduce the possibility of this occurring, the submitted message should be encoded in such a way that simple errors can be detected, making it possible in some cases to correct the error at the receiver end.

Each submitted symbol is encoded as a different codeword that is known in advance to both ends. When a codeword is obtained, the receiver searches for the most similar codeword using the Hamming distance. The coding is designed in such a way that the probability the receiver will misidentify the correct codeword is low. Peterson and Weldon (1972) noted that "much of binary coding theory has been based on the assumption that each symbol is affected independently by the noise, and therefore the probability of a given error pattern depends only on the number of errors."

Dietterich and Bariki (1995) employ communication techniques to transform multiclass problems into an ensemble of binary classification tasks. The idea is to transmit the correct class of a new instance via a channel composed of the instance attributes, the training data and the learning algorithm. Due to errors that may be present in the attributes, in the training data and/or failures in the classifier learning process, the class information can be disrupted. To provide the system with the ability to recover from these transmission errors, the class is codified by an error correcting code and each of its bits is transmitted separately, that is, through separate executions of the learning algorithm.

Accordingly, a distributed output code is used to represent the k classes associated to the multiclass problem. A codeword of length l is ascribed to each class. Commonly, the size of the codewords has more bits than needed in order to uniquely represent each class. The additional bits can be used to correct eventual classification errors. For this reason, the method is named error-correcting output coding (ECOC).

The generated codes are stored on a matrix $\vec{M} \in \{-1, +1\}^{k \times l}$. The rows of this matrix represent the codewords of each class and the columns correspond to the desired outputs of the l binary classifiers $(f_1(\vec{x}), \ldots, f_l(\vec{x}))$ induced.

A new pattern $\vec{x}$ can be classified by evaluating the classifications of the l classifiers, which generate a vector $\vec{f}(\vec{x})$ of length l. This vector is then measured against the rows of $\vec{M}$. The instance is ascribed to the class with the smallest Hamming distance.

In order to assess the goodness of ECOC, Dietterich and Bariki (1995) propose two criteria for row separation and column diversity:

Row separation. Codewords should be well-separated in Hamming distance;

Column diversity. Columns should be as uncorrelated as much as possible.

Dietterich and Bariki (1995) proposed that codewords should be designed in order to maximize their error-correcting capability and presented four techniques for constructing good error correcting codes. The choice of each technique is determined by the number of classes in the problem.

6.2.2 *Code-Matrix Framework*

Allwein *et al* (2000) presented a framework that which can be generally used to represent decomposition techniques. In this framework, the decomposition techniques are transformed into code-matrix based methods in which. a value from the set $\{-1, 0, +1\}$ is ascribed to each entry of the matrix $\vec{M}$. An entry m_{ij} with $+1$ value indicates that the class correspondent to row i assumes a positive label in the classifier f_j induction. The -1 value designates a negative label and the 0 value indicates that the data from class i does not participate in the classifier f_j induction. Binary classifiers are then trained to learn the labels represented in the columns of $\vec{M}$.

In the 1AA decomposition, $\vec{M}$ has dimension kxk, with diagonal entries equal to $+1$. All other entries receive the value -1. In the 1A1 case, $\vec{M}$ has dimension $kxk(k-1g)/2$ and each column corresponds to a binary classifier for a pair of classes (i, j). In each column representing a pair (i, j), the value of the entries corresponding to lines i and j are defined as $+1$ and -1, respectively. The remaining entries are equal to 0, indicating that patterns from the other classes do not participate in the induction of this particular binary classifier.

The classification of a new pattern's class involves a decoding step, as with EsCOC technique. Several decoding methods have been proposed in the literature [Passerini *et al.* (2004); Allwein *et al.* (2000); Windeatt and

Ghaderi (2003); Escalera *et al.* (2006); Klautau *et al.* (2003)].

No clear winner among various coding strategies such as 1AA, 1A1, dense random codes and sparse random codes, has been found in various experimental studies [Allwein *et al.* (2000)]. Thus, finding an adequate combination of binary classifiers for a given multiclass task can be considered a relevant research issue.

The next section presents the code-matrix design problem and discusses some of the main methods developed in this area. This problem can be defined as a search for codes to represent each class. Other issues to be addressed are the size of these codewords.

6.2.3   Code-matrix Design Problem

Several alternatives can be employed in order to decompose a multiclass problem into multiple binary subproblems. The most compact decomposition of a problem with k classes can be performed with the use of $l = \lceil \log_2(k) \rceil$ binary classifiers [Mayoraz and Moreira (1996)]. One instance of a compact matrix for a problem with four classes is presented in Table 6.5.

Table 6.5   Illustration of Compact Code-matrix for a Four-class Classification Problem.

Class Label	First Classifier	Second Classifier
Class 1	1	1
Class 2	1	-1
Class 3	-1	1
Class 4	-1	-1

The total number of different binary classifiers for a problem with k classes is $0.5\left(3^k + 1\right) - 2^k$, considering that $f = -f$. In other words, the inversion of the positive and negative labels produces the same classifier [Mayoraz and Moreira (1996)]. Among them, $2^{k-1} - 1$ include all classes simultaneously, i. e., they have only the labels $+1$ and -1, without the 0 entry. An example of a code-matrix constituted of such classifiers for a problem with four classes is illustrated in Table 6.6.

The next section reviews some strategies for obtaining ECOC matrices, i. e., code-matrices with an error-correcting capability and other strategies employed in obtaining code-matrices . Section 6.3 describes techniques for adapting code-matrices to each multiclass problem under consideration.

Table 6.6   Illustration of Compact Code-Matrix for a Four-Class Classification Problem.

Class Label	Classifier 1	Classifier 2	Classifier 3	Classifier 4	Classifier 5	Classifier 6
Class 1	1	1	1	1	1	1
Class 2	1	1	-1	-1	-1	-1
Class 3	-1	-1	1	1	-1	-1
Class 4	1	-1	1	-1	1	-1

Unless it is explicitly stated, the described works use binary code-matrices, that is, code-matrices with only $+1$ and -1 entries.

Dietterich and Bariki (1995) enforce two characteristics necessary to ensure error-correcting capability when designing ECOC matrices:

- Row separation;
- Column separation.

where the separation is measured through the Hamming distance, which is equal to the differences between different bit strings.

Constant columns (with only positive or negative entries) should also be avoided, since they do not represent a binary decision problem.

Let d_m designate the minimum Hamming distance between any pair of rows of $\vec{M}$. The final ECOC multiclass classifier is able to correct at least $\lfloor \frac{d_m-1}{2} \rfloor$ errors of the binary classifiers outputs. Since, according to the Hamming distance, each incorrect classification implies a deviation of one unity from the correct class codeword, committing $\lfloor \frac{d_m-1}{2} \rfloor$ errors, the closest codeword will still be that of the correct class [Dietterich and Bakiri (1995)]. This is the reason why a high row separation is demanded. According to this principle, the 1AA coding is unable to recover from any error, since its d_m is equal to 2. The row separation requirement is also demanded in designing error-correcting codes (ECC) in telecommunications [Alba and Chicano (2004)].

In addition to the above requirement, the errors of the binary classifiers that have been induced must be uncorrelated to obtain good error-correcting codes when solving the multiclass problem. In order to achieve this, column separation is also demanded, that is, the Hamming distance between each pair of columns of $\vec{M}$ must be high. If, in the learning algorithm, the inversion of the positive and negative labels produces the same classifier ($f = -f$), the Hamming distance between each column and the complement of the others must also be maximized.

Based on these observations, Dietterich and Bariki (1995) proposed four techniques for designing code-matrices with good error-correcting capability. The choice of each one is determined by the number of classes in the multiclass problem. No justification is given as to how the numbers of classes were stipulated for each method.

- For $k \leqslant 7$, it is recommended to use an exhaustive code. The first row (i.e. codeword of the first class) is composed of only $+1$ values. All other rows are composed of alternate runs of 2^{k-i} positive values and negative values when i is the row number. Table 6.5 illustrates the exhaustive code-matrix for the four-class problem.
- If $8 \leqslant k \leqslant 11$, a method that selects columns from the exhaustive code is applied.
- For $k > 11$, there are two options: a method based on the *hill-climbing* algorithm and the generation of BCH (Bose-Chaudhuri and Hocquenghem) codes [Boser and Ray-Chaudhuri (1960)]. The BCH algorithm employs polynomial functions to design nearly optimal error-correcting codes. One problem with BCH codes is that they do not ensure a good column separation. Moreover, if the number of classes is not a power of two, the code must be shortened by removing rows (and possibly columns) while maintaining good row and column separations.

Pimenta and Gama (2005) present an algorithm for designing ECOCs that obtain competitive predictive performance in relation to traditional decompositions, using decision trees (DTs) [Quinlan (1986)] and SVMs as base classifiers. They use d a function for evaluating the quality of ECOCs according to their error-correcting properties. An iterative persecution algorithm (PA) is used to construct the ECOCs by adding or removing columns from an initial ECOC in order to maximize the quality function.

A good ECOC is an ECOC that maximizes the minimum Hamming distance among codewords. Thus, they define a straight line $y = m.e + b$ where $n = \frac{2^{k-2}-1}{(2^{k-1}-1)-\lceil log_2(k)\rceil}$ and $b = 1 - m\lceil log_2(k)\rceil$. This line represents the best minimum Hamming distance for a certain k (number of classes) and e (the codeword size). Because Hamming distance obtain integer values, the support function $a(k,\ e)$ is defined as $y(k_1 e)$ rounded down. Based on this support function a Quality function $q(k, e)$ of an ECOC can be defined (assuming the values $W, B, B+$ comply with $W < B < B+$):

- $q(k, e) = W$ when the minimum Hamming distance of the ECOC is under the support function $a(k, e)$ at a distance greater than 1.

- $q(k, e) = B$ when the minimum Hamming distance of the ECOC is under the support function $a(k, e)$ at a distance lass than 1.
- $q(k, e) = B+$ when the minimum Hamming distance of the ECOC is in or over the support function $a(k, e)$.

Pimenta and Gama (2005) compared the performance of Repulsion Algorithm (RA) to the performance of Persecution Algorithm (PA). The RA tries to maximize an evaluation function that gets higher as d_m increases. Since row separation is not required in the design of an ECC, the evaluation function is used to penalize matrices with identical or complementary columns. Moreover, genetic algorithms (GAs) are used to design the code-matrices, for maximizing the evaluation function. The RA is used in the mutation step of the GA. The comparative study indicated that the PA performbetter in finding valid ECOCs, where the validity was measured by the criteria of avoiding equal, complementary and constant columns; the RA was the worst method. Among the valid ECOCs generated, the PA still performed, in general, better, obtaining ECOCs with good quality according to the evaluation function proposed by Pimenta and Gama (2005). Nevertheless, GARA (GA with RA) also designed ECOCs of good quality. Pimenta and Gama (2005) also suggested a method of determining the number of columns in the ECOC (i. e., the number of classifiers employed in the decomposition). This method involves examining an evaluation function based on the number of errors that can be corrected by ECOCs of different sizes.

There are some studies that claim that randomly designed ECOCs show good multiclass predictive performance [Berger (1999); Windeatt and Ghaderi (2003); Tapia *et al.* (2003)]. Allwein *et al* (2000) compared the use of two randomly designed matrices: dense and sparse. In the dense matrix, 10,000 random matrices were generated, with $\lceil 10 * \log_2(k) \rceil$ columns and entries receiving the values -1 or $+1$ with the same probability. The matrix with the higher d_m and without identical or complementary columns is chosen, following the recommendations of Dietterich and Bariki (1995) . In the sparse matrix, which uses the ternary alphabet, the number of columns in the code-matrix is $\lceil 15 \log_2 k \rceil$, and the entries are chosen as 0 with 0.5 probability and $+1$ or -1 with probability 0.25 each. Again, 10,000 random matrices are generated and the one with higher d_m is chosen.

Berger (1999) gives statistical and combinatorial arguments about why random matrices can perform well. Among these arguments, are theorems that state that random matrices are likely to show good row and column separation, specially as the number of columns increases. Nevertheless, it is assumed that the errors of the individual classifiers are uncorrelated, a state which does not hold for real applications.

Windeatt and Ghaderi (2002) also note the desirability of using equidistant codes. In equidistant codes the Hamming distance between rows is approximately constant. They showed that if $\vec{M}$ is an equidistant code-matrix, the number of +1's in different rows is the same and the number of common +1's between any pair of rows is equal. They used this heuristic to select a subset of rows from BCH codes, producing equidistant code-matrices. Experimentally, they verified that equidistant codes were superior to 1AA and random codes for shorter codes (with less columns), using multilayer perceptron (MLP) neural networks (NNs) [Haykin (1999)] as base classifiers. As the length of the codes increases, the coding strategy seems to be less significant, favoring a random design.

6.2.4 *Orthogonal Arrays (OA)*

In designing experiments, the aim is to minimize the number of experiments required to collect useful information about an unknown process (Montgomery, 1997). The collected data is typically used to construct a model for the unknown process. The model may then be used to optimize the original process.

A *full factorial design* is an experiment design in which the experimenter chooses n attributes that are believed to affect the target attribute. Then, all possible combinations of the selected input attributes are acquired [Montgomery (1997)]. Applying a full factorial design is impractical when there are many input attributes.

A *fractional factorial design* is a design in which only a fraction of the combinations required for the complete factorial experiment is selected. One of the most practical forms of fractional factorial design is the orthogonal array. An orthogonal array OA(n,k,d,t) is a matrix of k rows and n columns, with every entry being one of the d values. The array has strength t if, in every t by n submatrix, the d^t possible distinct rows all appear the same number of times. An example of a strength 2 OA is presented in Table 6.7. Any two classes in this array have all possible combinations ("1,1", "1,-1", "-1,1", "-1,-1"). Each of these combinations appears an equal number of times. In an orthogonal array of strength 3 presented in Table 6.8, we can find all combinations of any three classes.

Table 6.7 The OA(8,7,2,2) Design.

+1	+1	+1	+1	-1	-1	-1	-1
+1	+1	-1	-1	+1	+1	-1	-1
+1	+1	-1	-1	-1	-1	+1	+1
+1	-1	+1	-1	+1	-1	+1	-1
+1	-1	+1	-1	-1	+1	-1	+1
+1	-1	-1	+1	+1	-1	-1	+1
+1	-1	-1	+1	-1	+1	+1	-1

The columns, called runs of the OA matrix, represent tests which require resources such as time, money and hardware. OA aims to create the minimum number of runs to ensure the required strength. The advantage of compact representation of the OA is that we can use it to create the minimal number of binary classifiers.

Table 6.8 The OA(8,7,2,3) Design.

-1	-1	-1	-1	-1	-1	-1	-1	+1	+1	+1	+1	+1	+1	+1	+1
-1	+1	-1	+1	-1	+1	-1	+1	+1	-1	+1	-1	+1	-1	+1	-1
-1	-1	+1	+1	-1	-1	+1	+1	+1	+1	-1	-1	+1	+1	-1	-1
-1	+1	+1	-1	-1	+1	+1	-1	+1	-1	-1	+1	+1	-1	-1	+1
-1	-1	-1	-1	+1	+1	+1	+1	+1	+1	+1	+1	-1	-1	-1	-1
-1	+1	-1	+1	+1	-1	+1	-1	+1	-1	+1	-1	-1	+1	-1	+1
-1	-1	+1	+1	+1	+1	-1	-1	+1	+1	-1	-1	-1	-1	+1	+1
-1	+1	+1	-1	+1	-1	-1	+1	+1	-1	-1	+1	-1	+1	+1	-1

The number of rows k in the OA should be equal to the cardinality of the class set. The number of columns is equal to the number of classifiers that will be trained. Constructing a new OA design for any number of classes is not an easy task. Usually, it is possible to use a ready-made design for a specific number of classes. The orthogonal designs can be taken from Sloane's library of OAs (Sloane 2007). If no OA with the required number of rows k, can be found, then we can still take a OA with a larger number of rows and reduce it to the appropriate size. Note that any $N x k'$ subarray of an $OA(N, k, s, t)$ is an $OA(N, k', s, t')$, where $t' = min\{k', t\}$.

Before using the OA, we first need to remove columns that are the inverse of other columns, because the inversion of the positive and negative labels produces the same classifier [Mayoraz and Moreira (1996)]. For example columns 9 to 16 in Table 6.8 should be removed as they are an inversion of columns 1 to 8 respectively. Moreover, columns that are entirely labeled with +1 or −1 also need to be removed since there is no meaning to training a classifier with a single class. Accordingly, the first column 1 in Table 6.7 should be removed.

6.2.5 *Hadamard Matrix*

Zhang *et al* (2003) study the usefulness of Hadamard matrices for generating ECOCs in ensemble learning. Hadamard matrices may be regarded as special classes of two-level orthogonal arrays of strengths 2 and 3. These matrices are named after the French mathematician Jacques Hadamard (1865-1963).

They point out that these matrices can be considered optimal ECOCs, within the pool of k class codes that combine $k - 1$ base learners, where the optimality is measured according to the row and column separations criteria. Nevertheless, the Hadamard matrices are designed with numbers of rows of the power two. For other numbers of classes, some rows have to be deleted. Hadamard matrix provided a higher accuracy than random and 1AA matrices, when SVM algorithm is used as the binary classifiers induction.

A Hadamard matrix (HM) H_n consists of square matrices of $+1$'s and -1's whose rows are orthogonal which satisfies $H_n H_n^T = nI_n$. where $\mathbf{I}n$ is the nth order identity matrix. A Hadamard matrix $\mathbf{H}n$ is often written in the normalized form with both the first row and column consisting of all $+1$'s.

A Hadamard output code, obtained by removing the first column from any normalized Hadamard matrix, has two beneficial properties: every pair of codewords has the same Hamming distance; and b) every pair of columns is orthogonal.

6.2.6 *Probabilistic Error Correcting Output Code*

Up to this point we assumed that the binary classifiers were crisp classifiers and provided only the class label 1 or -1. However, most binary classifiers are probabilistic and thus provide the class distribution in addition to the class label. This distribution can be used to better select the appropriate class in case of ambiguous output code. For example rows 1, 2, 3, 5 and 8 in Table 6.2 have more than one predicted class.

We assume that the output of each classifier i is a 2-long vector $p_{i,1}(x), p_{i,2}(x)$. The values $p_{i,1}(x)$ and $p_{i,2}(x)$ represent the support that instance x belongs to class -1 and $+1$ respectively according to the classifier i. For the sake of simplicity, we assume the provided vector is the correct distribution i.e., $p_{i,1}(x) + p_{i,2}(x) = 1$. Kong and Dietterich (1995) use the following revised Hamming distance between the classifier outputs and the codeword of class j:

$$HD(x,j) = \sum_{i=1}^{l} \begin{cases} p_{i,2}(x) & \text{If } \tilde{M}_{j,i} = -1 \\ p_{i,1}(x) & \text{If } \tilde{M}_{j,i} = +1 \end{cases} \tag{6.1}$$

where $\vec{M}$ represents the code-matrix such as in Table 6.1. We now return to the decoding process presented in Table 6.2. However, we now assume that the classifier provides a classification distribution. Thus, instead of the classification outputs of $(-1, -1, -1)$ as in row 1 of Table 6.2 we obtain, for the sake of the example, the following distributions $0.8, 0.2$, $0.6, 0.4$ and $0.7, 0.3$ from classifiers 1 to 3 respectively. Note that these distributions correspond to the classification outputs of row 1. Using equation 6.1, we conclude that the distance from class Setosa (with codeword of $1, -1, 1$) is $0.8 + 0.4 + 0.7 = 1.9$. The distance from class Versicolor (with codeword of $-1, 1, 1$) is $0.2 + 0.4 + 0.7 = 1.3$. The distance from class Virginica (with codeword of $1, 1, -1$) is $0.8 + 0.4 + 0.3 = 1.5$. Therefore the selected class is Versicolor. Recall from Table 6.2 that if we use only the corresponding classification outputs of $(-1, -1, -1)$, any class could have been chosen (ambiguous case).

6.2.7 *Other ECOC Strategies*

This section presents code-matrix design works that could not be fully characterized into one of the classes described in the previous sections, either because they employ alternative criteria in the code-matrix design or because a combination of the error-correcting and adaptiveness criteria is used.

Sivalingam *et al.* [Sivalingam *et al.* (2005)] propose transforming a multiclass recognition problem into a minimal binary classification problem using the minimal classification method (MCM) aided by error-correcting codes. Instead of separating only two classes at each classification, MCM requires only $log_2 K$ classifications since this method separates two groups of multiple classes. Thus the MCM requires a small number of classifiers but can still provide similar accuracy performance to binary ECOC.

Mayoraz and Moreira (1996) introduce an iterative algorithm for code-matrix generation. The algorithm takes into account three criteria. The first two are the same criteria suggested by Dietterich and Bariki (1995). The third criterion is that each inserted column should be pertinent, according to the positions of the classes in the input space. A binary partition of classes is considered pertinent if it is easy to learn. The most important contribution of this algorithm is that the obtained classifiers are usually simpler than those induced by the original ECOC procedure.

Using concepts from telecommunications coding theory, Tapia *et al* (2001) present a particular class of ECOCs, recursive ECOCs (RECOC). The recursive codes are constructed from component subcodes of small lengths, which may be weak when working on their own, but strong when working together. This results in an ensemble of ECOCs, where each component subcode defines a local multiclass learner. Another interesting feature of RECOCs, noted by the authors, is that they allow a regulated degree of randomness in their design. Tapia *et al* (2003) indicate that a random code is the ideal way to protect information against noise.

As in channel coding theory, a puncturing mechanism can be used to prune the dependence among the binary classifiers in a code-matrix [Pérez-Cruz and Artés-Rodríguez (2002)]. This algorithm eliminates classifiers that degrade the performance of a previously designed code-matrix, deleting columns from it. As a result, less complex multiclass schemes can be obtained. In order to obtain these schemes, a ternary coding was employed, that is, the code-matrices could have positive, negative and null entries. Experimentally, they achieved a good performance when puncturing 1A1 and BCH ECOC codes.

Several attempts have been made to design code-matrices by maximizing certain diversity measures of classifier ensembles [Kuncheva (2005a)]. Specifically, the search for code combinations, in conjunction with the number of binary classifiers to compose the multiclass solution, constitute a combinatorial problem. In order to solve this combinatorial problem, we can employ genetic algorithms (GAs) [Mitchell (1999); Lorena and Carvalho (2008)]. The GAs are used to determine code-matrices according to: their accuracy performance; diversity measures among columns defined by Kuncheva (2005); or the margins of separation among codes of different classes [Shen and Tan (2005)]. Since the implemented GA also aims to minimize the ensemble size, the code-matrices are tailor-made for each multiclass problem.

6.3   Type II - Adapting Code-matrices to the Multiclass Problems

A common criticism of the 1AA, 1A1 and other ECOC strategies is that all of them perform the multiclass problem decomposition a priori, without taking into account the properties and characteristics of each application [Allwein *et al.* (2000); Mayoraz and Moreira (1996); Alpaydin and Mayoraz (1999); Mayoraz and Alpaydim (1998); Dekel and Singer (2003); Rätsch *et al.* (2003); Pujol *et al.* (2006)]. Furthermore, as Allwein *et al* (2000) point out, although the ECOC codes have good error-correcting properties, several of the binary subproblems created may be difficult to learn.

Data-driven error correcting output coding (DECOC) [Zhoua *et al.* (2008)]. explores the distribution of data classes and optimizes both the composition and number of base learners needed to design an effective and compact code matrix. Specifically, DECOC calculates the confidence score of each base classifier based on the structural information of the training data and use sorted confidence scores to assist in determining the code matrix of ECOC. The results show that the proposed DECOC is able to deliver competitive accuracy compared with other ECOC methods, using parsimonious base learners rather than the pairwise coupling (one-vs-one) decomposition scheme.

The key idea of DECOC is to reduce the number of learners by selectively including some of the binary learners into the code matrix. This optimizes both the composition and the number of base learners necessary to design an effective and compact code-matrix. Classifier selection is done by calculating a quality measure for each classifier which predicts how well each base-learner separates the training-set into two relatively homogenous groups. The main strength of DECOC is its ability to provide a competitive accuracy compared with other ECOC methods, using parsimonious base-learners. But this does not come for free. In the process of building, the DECOC uses cross-validation procedures. This process takes a lot of time, because in every iteration it builds a classifier for every base learner. This latter step is done as a pre-process and doesn't affect the running time for testing samples. But if there are time and space limitations for our pre-process with the training-set, it might be a problem.

Some attempts have been made to simultaneously find a code-matrix and the set of binary classifiers that produce the smallest empirical loss. Usually the columns of the code matrix and binary classifiers are created in a stage-wise manner.

Finding the optimal solution for this problem has been shown to be NP-hard [Crammer and Singer (2002)]. Nevertheless, several heuristics have been developed. The most popular heuristics attempt to combine the ECOC framework with the AdaBoost framework. More specifically, two algorithms are very widely used: output-code AdaBoost (AdaBoost.OC) [Schapire (1997)], and error-correcting code AdaBoost (AdaBoost.ECC) [Guruswami and Sahai (1999)].

Figure 6.2 presents the AdaBoost.OC algorithm. In each iteration, a weak classifier is induced. The instances are reweighted by focusing on the misclassified instances. Then, as in ECOC, a new binary decomposition (coloring) is selected in each iteration . There are several ways to find the coloring function. The simplest option is to uniformly and independently choose each value at random from $\{-1, 1\}$. A better option is to choose at random but to ensure an even split of the labels, i.e. half should be labeled as -1 and half as 1. The third option is to maximize the value of U_t by using optimization algorithms.

Given: $(x_1, y_1), \ldots, (x_m, y_m) : x_i \in X, y_i \in Y, |Y| = k$
Initialize: $\tilde{D}_1(i, l) = 1/m(k-1)$ if $l \neq y_i, \tilde{D}_1(l, l) = 0 \ \forall l \in Y$.

For $t = 1, 2, \ldots, T$:
   Compute coloring $\mu_t : Y \longrightarrow \{-1, +1\}$.
   Let $U_t = \sum_{i=1}^{m} \sum_{l \in Y} \tilde{D}_t(i, l) [\![\mu_t(y_i) \neq \mu_t(l)]\!]$.
   Let $D_t(i) = \frac{1}{U_t} \cdot \sum_{l \in Y} \tilde{D}_t(i, l) [\![\mu_t(y_i) \neq \mu_t(l)]\!]$.
   Get hypothesis $h_t : X \longrightarrow \{-1, +1\}$ from the weak learner for distribution D_t.
   Let $\tilde{h}_t(x) = \{l \in Y : h_t(x) = \mu_t(l)\}$.
   Let $\tilde{\varepsilon}_t = \frac{1}{2} \sum_{i=1}^{m} \sum_{l \in Y} \tilde{D}_t(i, l) \cdot ([\![y_i \notin \tilde{h}_t(x_i)]\!] + [\![l \in \tilde{h}_t(x_i)]\!])$.
   Let $\alpha_t = \frac{1}{2} \ln \left(\frac{1 - \tilde{\varepsilon}_t}{\tilde{\varepsilon}_t} \right)$.
   Update:
$$\tilde{D}_{t+1}(i, l) = \frac{1}{\tilde{Z}_t} \cdot \tilde{D}_t(i, l) \exp \left\{ \alpha_t([\![y_i \notin \tilde{h}_t(x_i)]\!] + [\![l \in \tilde{h}_t(x_i)]\!]) \right\}$$
   where $\tilde{Z}_t$ is a normalization factor.

Fig. 6.2   The algorithm AdaBoost.OC combining boosting and output coding.

Figure 6.3 presents the AdaBoost.ECC algorithm. The AdaBoost.ECC algorithm works similarly to AdaBoost.OC. However, instead of reweighting the instances based on the pseudo-loss, it uses the structure of the current classifier and its performance on the binary learning problem.

Sun *et al* (2005) prove that AdaBoost.ECC performs stage-wise functional gradient descent on a cost function, defined in the domain of margin values. Moreover Sun *et al* (2005) prove that AdaBoost.OC is a shrinkage version of AdaBoost.ECC. Shrinkage can be considered as a method, for improving accuracy in the case of noisy data. Thus, Sun *et al* (2005) conclude that in low noise datasets, AdaBoost.ECC may have some advantages over AdaBoost.OC. Yet, in noisy data sets, AdaBoost.OC should outperform AdaBoost.ECC.

Given: $(x_1, y_1), \ldots, (x_m, y_m) : x_i \in X, y_i \in Y, |Y| = k$
Initialize: $\tilde{D}_1(i, l) = 1/m(k - 1)$ if $l \neq y_i$, $\tilde{D}_1(i, l) = 0$ if $l = y_i$.

For $t = 1, 2, \ldots, T$:
  Compute coloring $\mu_t : Y \longrightarrow \{-1, +1\}$.
  Let $U_t = \sum_{i=1}^{m} \sum_{l \in Y} \tilde{D}_t(i, l) [\![\mu_t(y_i) \neq \mu_t(l)]\!]$.
  Let $D_t(i) = \frac{1}{U_t} \cdot \sum_{l \in Y} \tilde{D}_t(i, l) [\![\mu_t(y_i) \neq \mu_t(l)]\!]$.
  Get hypothesis $h_t : X \longrightarrow \{-1, +1\}$ from the weak learner for distribution D_t.
  Compute the weight of positive and negative votes α_t and β_t respectively.
  Define:
    $g_t(x) = \alpha_t$ if $h_t(x) = +1$
    $\quad\quad = -\beta_t$ if $h_t(x) = -1$.
  Update:
    $\tilde{D}_{t+1}(i, l) = \frac{1}{\tilde{Z}_t} \cdot \tilde{D}_t(i, l) \exp\left\{ (g_t(x_i)\mu_t(l) - g_t(x_i)\mu_t(y_i))/2 \right\}$
  where $\tilde{Z}_t$ is a normalization factor.

Fig. 6.3 The algorithm AdaBoost.ECC combining boosting and output coding.

Crammer and Singer (2000) explain how to design code-matrix by adapting it to each multiclass problem under consideration. Finding a discrete code-matrix can be considered a NP-hard problem. Thus, they relaxed the problem, by allowing the matrix elements having continuous values. Thereupon, they obtained a variant of SVMs for the direct solution of multiclass problems. The predictive performance of this technique is comparable to those of the 1AA and 1A1 strategies [Hsu and Lin (2002)]. Nevertheless, the computational cost of this adaptive training algorithm is higher than the ready made matrices.

One can also combine linear binary classifiers in order to obtain a nonlinear multiclass classifier [Alpaydin and Mayoraz (1999)] . In this process, a MLP NN is obtained in which the first weight layer represents the parameters of the linear classifiers; the internal nodes correspond to the linear classifiers; and the final weight layer is equivalent to the code-matrix. This NN has an architecture and second layer weights initialized according to a code-matrix. As a result, the code-matrix and classifiers parameters are optimized jointly in the NN training. The proposed method showed higher accuracy than those of 1AA, 1A1 and ECOC decompositions employing linear binary classifiers.

Dekel and Singer (2003) develop a bunching algorithm which adapts code-matrices to the multiclass problem during the learning process. First, the training instances are mapped to a common space where it is possible to measure the divergence between input instances and their labels. Two matrices are used in the mapping process, one for the data and the other for the labels, which is the code-matrix. These two matrices are iteratively adapted by the algorithm in order to obtain a minimum error for the training data. This error is measured by the divergence between the training data and their labels in the common space. The code-matrices are probabilistic. Given an initial code-matrix, the the algorithm modifies it according to the previous procedure. Given a new instance, it is mapped to the new space and the predicted class is the one closer to the instance in this space. This algorithm was used for improving the performance of logistic regression classifiers [Collins *et al.* (2002)].

Rätsch *et al* (2003) define an optimization problem, in which the codes and the embedding functions f are determined jointly by maximizing a margin measure. The embedding functions ensure that instances from the same class are close to their respective codeword vector. The margin is defined as the difference between the distance of $\vec{f}(\vec{x})$ from the actual class and the to closer incorrect class.

In [Pujol *et al.* (2006)] a heuristic method for designing ternary code-matrices is introduced. The design is based on a hierarchical partition of the classes according to a discriminative criterion. The criterion used was the mutual information between the feature data and its class label. Initiating with all classes, they are recursively partitioned into two subsets in order to maximize the mutual information measure until each subset contains one class. These partitions define the binary classifiers to be employed in the code-matrix. For a problem with k classes, $k - 1$ binary classifiers are generated in this process. Experimental results demonstrated the potential of the approach using DTs and boosted decision stumps (BDS) [Freund and Schapire (1997)] as base classifiers. The algorithm showed results that were competitive to 1AA, 1A1 and random code-matrices.

In [Lorena and Carvalho (2007)], GAs were used to determine ternary code-matrices according to their performance in multiclass problem solution. In such cases, the code-matrices are adapted to each multiclass problem. Another goal in implementing GAs is to minimize the number of columns in the matrices in order to produce simpler decompositions.

Chapter 7

Evaluating Ensembles of Classifiers

In this chapter we introduce the main concepts and quality criteria in classifiers ensemble evaluation.

Evaluating the performance of an ensemble is a fundamental aspect of pattern recognition. The evaluation is important for understanding the quality of a certain ensemble algorithm and for tuning its parameters.

While there are several criteria for evaluating the predictive performance of ensemble of classifiers, other criteria such as the computational complexity or the comprehensibility of the generated ensemble can be important as well.

7.1 Generalization Error

Historically predictive performance measures are the main criteria for selecting inducers. Moreover, the predictive performance measures, such as accuracy, are considered to be an objective and quantified, which can be easily used to benchmark algorithms.

In addition to the experimental studies that are performed to validate the contribution of a new specific ensemble method, there are several large comparative studies, which aim to assist the practitioner in his decision making.

Let $E(S)$ represent an ensemble trained on dataset S. The generalization error of $E(S)$ is its probability to misclassify an instance selected according to the distribution D of the labeled instance space. The *classification accuracy* of an ensemble is one minus the generalization error. The *training error* is defined as the percentage of examples in the training set correctly classified by the ensemble, formally:

$$\hat{\varepsilon}(E(S), S) = \sum_{\langle x,y \rangle \in S} L\left(y, E(S)(x)\right) \tag{7.1}$$

where $L(y, E(S)(x))$ is the zero-one loss function defined in Equation 1.3.

In this book, classification accuracy is the primary evaluation criterion.

Although generalization error is a natural criterion, its actual value is known only in rare cases (mainly synthetic cases). The reason for that is that the distribution D of the labeled instance space is not known.

One can take the training error as an estimation of the generalization error. However, using the training error as is will typically provide an optimistically biased estimate, especially if the inducer *over-fits* the training data. There are two main approaches for estimating the generalization error: Theoretical and Empirical. In this book we utilize both approaches.

7.1.1   *Theoretical Estimation of Generalization Error*

A low training error does not guarantee low generalization error. There is often a trade-off between the training error and the confidence assigned to the training error as a predictor for the generalization error, measured by the difference between the generalization and training errors. The capacity of the inducer is a major factor in determining the degree of confidence in the training error. In general, the capacity of an inducer indicates the variety of classifiers it can induce. Breiman's upper bound on the generalization error of random forest (Breiman, 2001) which is expressed "in terms of two parameters that are measures of how accurate the individual classifiers are and of the agreement between them". While Breiman's bound is theoretically justified, it is not considered to be very tight (Kuncheva, 2004). Bartlett and Traskin (2007) showed that AdaBoost is almost surely consistent (i.e. ensemble's risk converges to the Bayes risk), if stopped sufficiently early, after $m^{1-\alpha}$ iterations where m is the training set size. However they could not determine whether this number can be increased.

Large ensemble with many members, relative to the size of the training set, are likely to obtain a low training error. On the other hand, they might just be memorizing or overfitting the patterns and hence exhibit a poor generalization ability. In such cases, the low error is likely to be a poor predictor of the higher generalization error. When the opposite occurs, that is to say, when capacity is too small for the given number of examples, inducers may underfit the data, and exhibit both poor training and generalization error.

In "Mathematics of Generalization", [Wolpert (1995)] discuss four theoretical frameworks for estimating the generalization error: Probably Approximately Correct (PAC), VC and Bayesian, and statistical physics. All these frameworks combine the training error (which can be easily calculated) with some penalty function expressing the capacity of the inducers.

7.1.2 *Empirical Estimation of Generalization Error*

Another approach for estimating the generalization error is the holdout method in which the given dataset is randomly partitioned into two sets: training and test sets. Usually, two-thirds of the data is considered for the training set and the remaining data are allocated to the test set. First, the training set is used by the inducer to construct a suitable classifier and then we measure the misclassification rate of this classifier on the test set. This test set error usually provides a better estimation of the generalization error than the training error. The reason for this is the fact that the training error usually under-estimates the generalization error (due to the overfitting phenomena). Nevertheless since only a proportion of the data is used to derive the model, the estimate of accuracy tends to be pessimistic.

A variation of the holdout method can be used when data is limited. It is common practice to resample the data, that is, partition the data into training and test sets in different ways. An inducer is trained and tested for each partition and the accuracies averaged. By doing this, a more reliable estimate of the true generalization error of the inducer is provided.

Random subsampling and n-fold cross-validation are two common methods of resampling. In random subsampling, the data is randomly partitioned several times into disjoint training and test sets. Errors obtained from each partition are averaged. In n-fold cross-validation, the data is randomly split into n mutually exclusive subsets of approximately equal size. An inducer is trained and tested n times; each time it is tested on one of the k folds and trained using the remaining $n-1$ folds.

The cross-validation estimate of the generalization error is the overall number of misclassifications divided by the number of examples in the data. The random subsampling method has the advantage that it can be repeated an indefinite number of times. However, a disadvantage is that the test sets are not independently drawn with respect to the underlying distribution of examples. Because of this, using a t-test for paired differences with random subsampling can lead to an increased chance of type I error, i.e., identifying a significant difference when one does not actually exist. Using a t-test on the generalization error produced on each fold lowers the chances of type I error but may not give a stable estimate of the generalization error. It is common practice to repeat n-fold cross-validation n times in order to provide a stable estimate. However, this, of course, renders the test sets non-independent and increases the chance of type I error. Unfortunately, there is no satisfactory solution to this problem. Alternative tests suggested by [Dietterich (1998)] have a low probability of type I error but a higher chance of type II error that is, failing to identify a significant difference when one does actually exist.

Stratification is a process often applied during random subsampling and n-fold cross-validation. Stratification ensures that the class distribution from the whole dataset is preserved in the training and test sets. Stratification has been shown to help reduce the variance of the estimated error especially for datasets with many classes.

Another cross-validation variation is the bootstraping method which is a n-fold cross validation, with n set to the number of initial samples. It samples the training instances uniformly with replacement and leave-one-out. In each iteration, the classifier is trained on the set of $n - 1$ samples that is randomly selected from the set of initial samples, S. The testing is performed using the remaining subset.

Dietterich [Dietterich (2000a)] has compared three methods for constructing forest of C4.5 classifiers: Randomizing, Bagging , and Boosting. The experiments show that when there is little noise in the data, boosting gives the best results. Bagging and Randomizing are usually equivalent. Another study [Bauer and Kohavi (1999)] compared Bagging and Boosting using decision trees and naive Bayes. The study determines that Bagging reduced variance of unstable methods, while boosting methods reduced both the bias and variance of unstable methods but increased the variance for stable methods.

Additional study [Opitz and Maclin (1999)] that compared Bagging with Boosting using neural networks and decision trees indicates that Bagging is sometimes significantly less accurate than Boosting. The study indicates that the performance of the Boosting methods is much more sensitive to the characteristics of the dataset, specifically Boosting may overfit noisy data sets and reducing classification performance.

Villalba *et al.* (2003) compared seven different boosting methods. They conclude that for binary classification tasks - the well-known AdaBoost should be preferred. However for multi-class tasks other boosting methods such as GentleAdaBoost should be considered.

A recent research has experimentally evaluated bagging and seven other randomization-based approaches for creating an ensemble of decision tree classifiers [Banfield *et al.* (2007)]. Statistical tests were performed on experimental results from 57 publicly available datasets. When cross-validation comparisons were tested for statistical significance, the best method was statistically more accurate than bagging on only eight of the 57 datasets. Alternatively, examining the average ranks of the algorithms across the group of datasets, Banfield *et al* found that boosting, random forests, and randomized trees is statistically significantly better than bagging.

Sohna and Shinb [Sohna (2007)] compared the performance of several ensemble methods (bagging, modified random subspace method, classifier selection, parametric fusion) to a single classifier assuming that the base inducer is logistic regression. They argue that several factors should be taken into consideration when performing such comparison, including correlation between input variables; variance of observation, and training data set size. They show that for large training sets, the performances of a single logistic regression and bagging are not significantly different. However, when training set size are small, bagging is superior to a single logistic regression classifier. When training data set size is small and correlation is strong, both modified random subspace method and bagging perform better than the other methods. When correlation is weak and variance is small, both parametric fusion and classifier selection algorithm appear to be the worst.

7.1.3 *Alternatives to the Accuracy Measure*

Accuracy is not a sufficient measure for evaluating a model with an imbalanced distribution of the class. There are cases where the estimation of an accuracy rate may mislead one about the quality of a derived classifier. In such circumstances, where the dataset contains significantly more majority class than minority class instances, one can always select the majority class and obtain good accuracy performance. Therefore, in these cases, the sensitivity and specificity measures can be used as an alternative to the accuracy measures [Han and Kamber (2001)].

Sensitivity (also known as recall) assesses how well the classifier can recognize positive samples and is defined as

$$Sensitivity = \frac{true\_positive}{positive} \tag{7.2}$$

where *true_positive* corresponds to the number of the true positive samples and positive is the number of positive samples.

Specificity measures how well the classifier can recognize negative samples. It is defined as

$$Specificity = \frac{true\_negative}{negative} \tag{7.3}$$

where *true_negative* corresponds to the number of the true negative examples and negative the number of samples that is negative.

Another well-known performance measure is precision. Precision measures how many examples classified as "positive" class are indeed "positive". This measure is useful for evaluating crisp classifiers that are used to classify an entire dataset. Formally:

$$Precision = \frac{true\_positive}{true\_positive + false\_positive} \tag{7.4}$$

Based on the above definitions the *accuracy* can be defined as a function of *sensitivity* and *specificity*:

$$Accuracy = Sensitivity \cdot \frac{positive}{positive+negative} + \\ Specificity \cdot \frac{negative}{positive+negative} \tag{7.5}$$

7.1.4 The F-Measure

Usually there is a tradeoff between the precision and recall measures. Trying to improve one measure often results in a deterioration of the second measure. Figure 7.1 illustrates a typical precision-recall graph. This two-dimensional graph is closely related to the well-known receiver operating characteristics (ROC) graphs in which the true positive rate (recall) is plotted on the Y-axis and the false positive rate is plotted on the X-axis [Ferri *et al.* (2002)]. However unlike the precision-recall graph, the ROC diagram is always convex.

Given a probabilistic classifier, this trade-off graph may be obtained by setting different threshold values. In a binary classification problem, the classifier prefers the class "not pass" over the class "pass" if the probability

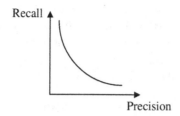

Fig. 7.1   A typical precision-recall diagram.

for "not pass" is at least 0.5. However, by setting a different threshold value other than 0.5, the trade-off graph can be obtained.

The problem here is described as multi-criteria decision-making (MCDM). The simplest and the most commonly used method to solve MCDM is the weighted sum model. This technique combines the criteria into a single value by using appropriate weighting. The basic principle behind this technique is the additive utility assumption. The criteria measures must be numerical, comparable and expressed in the same unit. Nevertheless, in the case discussed here, the arithmetic mean can mislead. Instead, the harmonic mean provides a better notion of "average". More specifically, this measure is defined as [Van Rijsbergen (1979)]:

$$F = \frac{2 \cdot P \cdot R}{P + R} \tag{7.6}$$

The intuition behind the F-measure can be explained using Figure 7.2. Figure 7.2 presents a diagram of a common situation in which the right ellipsoid represents the set of all defective batches and the left ellipsoid represents the set of all batches that were classified as defective by a certain classifier. The intersection of these sets represents the true positive (TP), while the remaining parts represent false negative (FN) and false positive (FP). An intuitive way of measuring the adequacy of a certain classifier is to measure to what extent the two sets match, namely to measure the size of the unshaded area. Since the absolute size is not meaningful, it should be normalized by calculating the proportional area. This value is the F-measure:

Proportion of unshaded area =

$$\frac{2 \cdot (True\ Positive)}{False\ Positive + False\ Negative + 2 \cdot (True\ Positve)} = F \tag{7.7}$$

The F-measure can have values between 0 to 1. It obtains its highest value when the two sets presented in Figure 7.2 are identical and it obtains its lowest value when the two sets are mutually exclusive. Note that each point on the precision-recall curve may have a different F-measure. Furthermore, different classifiers have different precision-recall graphs.

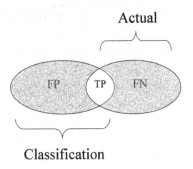

Fig. 7.2   A graphic explanation of the F-measure.

7.1.5   *Confusion Matrix*

The confusion matrix is used as an indication of the properties of a classification (discriminant) rule. It contains the number of elements that have been correctly or incorrectly classified for each class. We can see on its main diagonal the number of observations that have been correctly classified for each class; the off-diagonal elements indicate the number of observations that have been incorrectly classified. One benefit of a confusion matrix is that it is easy to see if the system is confusing two classes (i.e. commonly mislabelling one as an other).

For every instance in the test set, we compare the actual class to the class that was assigned by the trained classifier. A positive (negative) example that is correctly classified by the classifier is called a true positive (true negative); a positive (negative) example that is incorrectly classified is called a false negative (false positive). These numbers can be organized in a confusion matrix shown in Table 7.1.

Based on the values in Table 7.1, one can calculate all the measures defined above:

- Accuracy is: $(a+d)/(a+b+c+d)$
- Misclassification rate is: $(b+c)/(a+b+c+d)$

Table 7.1   A Confusion Matrix.

	Predicted negative	Predicted positive
Negative Examples	A	B
Positive Examples	C	D

- Precision is: $d/(b + d)$
- True positive rate (Recall) is: $d/(c + d)$
- False positive rate is: $b/(a + b)$
- True negative rate (Specificity) is: $a/(a + b)$
- False negative rate is: $c/(c + d)$

7.1.6   *Classifier Evaluation under Limited Resources*

The above mentioned evaluation measures are insufficient when probabilistic classifiers are used for choosing objects to be included in a limited quota. This is a common situation that arises in real-life applications due to resource limitations that require cost-benefit considerations. Resource limitations prevent the organization from choosing all the instances. For example, in direct marketing applications, instead of mailing everybody on the list, the marketing efforts must implement a limited quota, i.e., target the mailing audience with the highest probability of positively responding to the marketing offer without exceeding the marketing budget.

Another example deals with a security officer in an air terminal. Following September 11, the security officer needs to search all passengers who may be carrying dangerous instruments (such as scissors, penknives and shaving blades). For this purpose the officer is using a classifier that is capable of classifying each passenger either as class A, which means, "Carry dangerous instruments" or as class B, "Safe".

Suppose that searching a passenger is a time-consuming task and that the security officer is capable of checking only 20 passengers prior to each flight. If the classifier has labeled exactly 20 passengers as class A, then the officer will check all these passengers. However if the classifier has labeled more than 20 passengers as class A, then the officer is required to decide which class A passenger should be ignored. On the other hand, if less than 20 people were classified as A, the officer, who must work constantly, has to decide who to check from those classified as B after he has finished with the class A passengers.

There also cases in which a quota limitation is known to exist but its size is not known in advance. Nevertheless, the decision maker would like to evaluate the expected performance of the classifier. Such cases occur, for example, in some countries regarding the number of undergraduate students that can be accepted to a certain department in a state university. The actual quota for a given year is set according to different parameters including governmental budget. In this case, the decision maker would like to evaluate several classifiers for selecting the applicants while not knowing the actual quota size. Finding the most appropriate classifier in advance is important because the chosen classifier can dictate what the important attributes are, i.e. the information that the applicant should provide the registration and admission unit.

In probabilistic classifiers, the above mentioned definitions of precision and recall can be extended and defined as a function of a probability threshold τ . If we evaluate a classifier based on a given a test set which consists of n instances denoted as $(< x_1, y_1 >, \ldots, < x_n, y_n >)$ such that x_i represents the input features vector of instance i and y_i represents its true class ("positive" or "negative"), then:

$$\text{Precision } (\tau) = \frac{\left| \{< x_i, y_i >: \hat{P}_E(pos \,|x_i) > \tau, y_i = pos\} \right|}{\left| \{< x_i, y_i >: \hat{P}_E(pos \,|x_i) > \tau \right|} \qquad (7.8)$$

$$\text{Recall } (\tau) = \frac{\left| \{< x_i, y_i >: \hat{P}_E(pos \,|x_i) > \tau, y_i = pos\} \right|}{|\{< x_i, y_i >: y_i = pos\}|} \qquad (7.9)$$

where E represents a probabilistic ensemble that is used to estimate the conditional likelihood of an observation x_i to "positive" which is denoted as $\hat{P}_E(pos \,|x_i)$. The typical threshold value of 0.5 means that the predicted probability of "positive" must be higher than 0.5 for the instance to be predicted as "positive". By changing the value of τ, one can control the number of instances that are classified as "positive". Thus, the τ value can be tuned to the required quota size. Nevertheless because there might be several instances with the same conditional probability, the quota size is not necessarily incremented by one.

The above discussion is based on the assumption that the classification problem is binary. In cases where there are more than two classes, adaptation could be easily made by comparing one class to all the others.

7.1.6.1 *ROC Curves*

Another measure is the receiver operating characteristic (ROC) curves which illustrate the tradeoff between true positive to false positive rates [Provost and Fawcett (1998)]. Figure 7.3 illustrates a ROC curve in which the X-axis represents a false positive rate and the Y-axis represents a true positive rate. The ideal point on the ROC curve would be (0,100), that is, all positive examples are classified correctly and no negative examples are misclassified as positive.

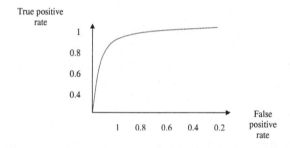

Fig. 7.3 A typical ROC curve.

The ROC convex hull can also be used as a robust method of identifying potentially optimal classifiers [Provost and Fawcett (2001)]. Given a family of ROC curves, the ROC convex hull can include points that are more towards the north-west frontier of the ROC space. If a line passes through a point on the convex hull, then there is no other line with the same slope passing through another point with a larger true positive (TP) intercept. Thus, the classifier at that point is optimal under any distribution assumptions in tandem with that slope.

7.1.6.2 *Hit Rate Curve*

The hit rate curve presents the hit ratio as a function of the quota size[An and Wang (2001)]. *Hit rate* is calculated by counting the actual positive labeled instances inside a determined quota. More precisely for a quota of size j and a ranked set of instances, *hit rate* is defined as:

$$\text{HitRate}(j) = \frac{\sum_{k=1}^{j} t^{[k]}}{j} \qquad (7.10)$$

where $t^{[k]}$ represents the truly expected outcome of the instance located in the k'th position when the instances are sorted according to their conditional probability for "positive" by descending order. Note that if the k'th position can be uniquely defined (i.e. there is exactly one instance that can be located in this position) then $t^{[k]}$ is either 0 or 1 depending on the actual outcome of this specific instance. Nevertheless if the k'th position is not uniquely defined and there are $m_{k,1}$ instances that can be located in this position, and $m_{k,2}$ of which are truly positive, then:

$$t^{[k]} = m_{k,2}/m_{k,1} \qquad (7.11)$$

The sum of $t^{[k]}$ over the entire test set is equal to the number of instances that are labeled "positive". Moreover $Hit - Rate(j) \approx Precision(p^{[j]})$ where $p^{[j]}$ denotes the j'th order of $\hat{P}_I(pos\,|x_1), \cdot, \hat{P}_I(pos\,|x_m)$. The values are strictly equal when the value of j'th is uniquely defined.

7.1.6.3   Qrecall (Quota Recall)

The hit-rate measure, presented above, is the "precision" equivalent for quota-limited problems. Similarly, we suggest the Qrecall (for quota recall) to be the "recall" equivalent for quota-limited problems. The Qrecall for a certain position in a ranked list is calculated by dividing the number of positive instances, from the head of the list until that position, by the total positive instances in the entire dataset. Thus, the Qrecall for a quota of j is defined as:

$$\text{Qrecall}(j) = \frac{\sum_{k=1}^{j} t^{[k]}}{n^+} \qquad (7.12)$$

The denominator stands for the total number of instances that are classified as positive in the entire dataset. Formally it can be calculated as:

$$n^+ = |\{< x_i, y_i >: y_i = pos\}| \qquad (7.13)$$

7.1.6.4   Lift Curve

A popular method of evaluating probabilistic models is *lift* [Coppock (2002)]. After a ranked test set is divided into several portions (usually deciles), lift is calculated as follows: the ratio of really positive instances in a specific decile is divided by the average ratio of really positive instances in the population. Regardless of how the test set is divided, a good model is

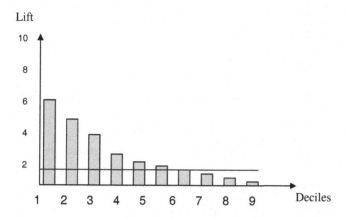

Fig. 7.4 A typical lift chart.

achieved if the lift decreases when proceeding to the bottom of the scoring list. A good model would present a lift greater than 1 in the top deciles and a lift smaller than 1 in the last deciles. Figure 7.4 illustrates a lift chart for a typical model prediction. A comparison between models can be done by comparing the lift of the top portions, depending on the resources available and cost/benefit considerations.

7.1.6.5 *Pearson Correlation Coefficient*

There are also some statistical measures that may be used as performance evaluators of models. These measures are well-known and can be found in many statistical books. In this section we examine the Pearson correlation coefficient. This measure can be used to find the correlation between the ordered estimated conditional probability $(p^{[k]})$ and the ordered actual expected outcome $(t^{[k]})$. A Pearson correlation coefficient can have any value between -1 and 1 where the value 1 represents the strongest positive correlation. It should be noticed that this measure take into account not only the ordinal place of an instance but also its value (i.e. the estimated probability attached to it). The *Pearson* correlation coefficient for two random variables is calculated by dividing the co-variance by the product of both standard deviations. In this case, the standard deviations of the two variables assuming a quota size of j are:

$$\sigma_p(j) = \sqrt{\frac{1}{j}\sum_{i=1}^{j}\left(p^{[i]} - \bar{p}(j)\right)} \; ; \; \sigma_t(j) = \sqrt{\frac{1}{j}\sum_{i=1}^{j}\left(t^{[i]} - \bar{t}(j)\right)} \tag{7.14}$$

where $\bar{p}(j), \bar{t}(j)$ represent the average of $p^{[i]}$'s and $t^{[i]}$'s respectively:

$$\bar{p}(j) = \frac{\sum_{i=1}^{j} p^{[i]}}{j} \quad ; \quad \bar{t}(j) = \frac{\sum_{i=1}^{j} t^{[i]}}{j} = HitRate(j) \tag{7.15}$$

The co-variance is calculated as follows:

$$Cov_{p,t}(j) = \frac{1}{j}\sum_{i-1}^{j}\left(p^{[i]} - \bar{p}(j)\right)\left(t^{[i]} - \bar{t}(j)\right) \tag{7.16}$$

Thus, the Pearson correlation coefficient for a quota j, is:

$$\rho_{p,t}(j) = \frac{Cov_{p,t}(j)}{\sigma_p(j) \cdot \sigma_t(j)} \tag{7.17}$$

7.1.6.6　*Area Under Curve (AUC)*

Evaluating a probabilistic model without using a specific fixed quota is not a trivial task. Using continuous measures like hit curves, ROC curves and lift charts, mentioned previously, is problematic. Such measures can give a definite answer to the question "Which is the best model?" only if one model dominates in the curve space, meaning that the curves of all the other model are beneath it or equal to it over the entire chart space. If a dominating model does not exist, then there is no answer to that question, using only the continuous measures mentioned above.. Complete order demands no intersections of the curves. Of course, in practice there is almost never one dominating model. The best answer that can be obtained is in regard to which areas one model outperforms the others. As shown in Figure 7.5, every model gets different values in different areas. If a complete order of model performance is needed, another measure should be used.

Area under the ROC curve (AUC) is a useful metric for classifier performance since it is independent of the decision criterion selected and prior probabilities. The AUC comparison can establish a dominance relationship between classifiers. If the ROC curves are intersecting, the total AUC is an average comparison between models [Lee (2000)]. The bigger it is, the better the model is. As opposed to other measures, the area under the

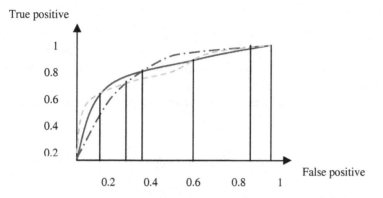

Fig. 7.5 Areas of dominancy. A ROC curve is an example of a measure that gives areas of dominancy and not a complete order of the models. In this example the equally dashed line model is the best for f.p (false positive) < 0.2. The full line model is the best for 0.2 < f.p <0.4. The dotted line model is best for 0.4 < f.p < 0.9 and from 0.9 to 1 again the dashed line model is the best.

ROC curve (AUC) does not depend on the imbalance of the training set [Kolcz (2003)]. Thus, the comparison of the AUC of two classifiers is fairer and more informative than comparing their misclassification rates.

7.1.6.7 *Average Hit Rate*

The average hit rate is a weighted average of all hit-rate values. If the model is optimal, then all the really positive instances are located in the head of the ranked list, and the value of the average hit rate is 1. The use of this measure fits an organization that needs to minimize type II statistical error (namely, to include a certain object in the quota although in fact this object will be labeled as "negative"). Formally the Average Hit Rate for binary classification problems is defined as:

$$AverageHitRate = \frac{\sum_{j=1}^{n} t^{[j]} \cdot HitRate(j)}{n^+} \tag{7.18}$$

where $t^{[j]}$ is defined as in Equation 4 and is used as a weighting factor. Note that the average hit rate ignores all hit rate values on unique positions that are actually labeled as "negative" class (because $t^{[j]}=0$ in these cases).

7.1.6.8 *Average Qrecall*

Average Qrecall is the average of all the Qrecalls which extends from the position that is equal to the number of positive instances in the test set to the bottom of the list. Average Qrecall fits an organization that needs to minimize type I statistical error (namely, not including a certain object in the quota although in fact this object will be labeled as "positive"). Formally, average Qrecall is defined as:

$$\frac{\sum\limits_{j=n^+}^{n} Qrecall(j)}{n - (n^+ - 1)} \tag{7.19}$$

where n is the total number of instances and n^+ is defined in Equation (7.13).

Table 7.2 illustrates the calculation of average Qrecall and average hit-rate for a dataset of ten instances. The table presents a list of instances in descending order according to their predicted conditional probability to be classified as "positive". Because all probabilities are unique, the third column ($t^{[k]}$) indicates the actual class ("1" represents "positive" and "0" represents "negative"). The average values are simple algebraic averages of the highlighted cells.

Table 7.2 An Example for Calculating Average Qrecall and Average Hit-rate.

Place in list (j)	Positive probability	$t^{[k]}$	Qrecall	Hit rate
1	0.45	1	0.25	1
2	0.34	0	0.25	0.5
3	0.32	1	0.5	0.667
4	0.26	1	0.75	0.75
5	0.15	0	0.75	0.6
6	0.14	0	0.75	0.5
7	0.09	1	1	0.571
8	0.07	0	1	0.5
9	0.06	0	1	0.444
10	0.03	0	1	0.4
		Average:	0.893	0.747

Note that both *average Qrecall* and *average hit rate* get the value 1 in an optimum classification, where all the positive instances are located at the head of the list. This case is illustrated in Table 7.3. A summary of the key differences are provided in Table 7.4.

Table 7.3   Qrecall and Hit-rate in an Optimum Prediction.

Place in list (j)	Positive probability	$t^{[k]}$	Qrecall	Hit rate
1	0.45	1	0.25	1
2	0.34	1	0. 5	1
3	0.32	1	0.75	1
4	0.26	1	1	1
5	0.15	0	1	0.8
6	0.14	0	1	0.667
7	0.09	0	1	0.571
8	0.07	0	1	0.5
9	0.06	0	1	0.444
10	0.03	0	1	0.4
		Average:	1	1

Table 7.4   Characteristics of Qrecall and Hit-rate.

Parameter	Hit-rate	Qrecall
Function increasing/decreasing	Non monotonic	Monotonically increasing
End point	Proportion of positive samples in the set	1
Sensitivity of the measures value to positive instances	Very sensitive to positive instances at the top of the list. Less sensitive on going down to the bottom of the list.	Same sensitivity to positive instances in all places in the list.
Effect of negative class on the measure	A negative instance affects the measure and cause its value to decrease.	A negative instance does not affect the measure.
Range	$0 \leq$ Hit-rate ≤ 1	$0 \leq$ Qrecall ≤ 1

7.1.6.9   *Potential Extract Measure (PEM)*

To better understand the behavior of Qrecall curves, consider the cases of random prediction and optimum prediction.

Suppose no learning process was applied on the data and the list produced as a prediction would be the test set in its original (random) order. On the assumption that positive instances are distributed uniformly in the population, then a quota of random size contains a number of positive instances that are proportional to the a-priori proportion of positive instances in the population. Thus, a Qrecall curve that describes a uniform distribution (which can be considered as a model that predicts as well as a random guess, without any learning) is a linear line (or semi-linear because values are discrete) which starts at 0 (for zero quota size) and ends in 1.

Suppose now that a model gave an optimum prediction, meaning that all positive instances are located at the head of the list and below them, all the negative instances. In this case, the Qrecall curve climbs linearly until a value of 1 is achieved at point n^+ (n^+ = number of positive samples). From that point any quota that has a size bigger than n^+, fully extracts test set potential and the value 1 is kept until the end of the list.

Note that a "good model", which outperforms random classification, though not an optimum one, will fall "on average" between these two curves. It may drop sometimes below the random curve but generally, more area is delineated between the "good model" curve and the random curve, above the latter than below it. If the opposite is true then the model is a "bad model" that does worse than a random guess.

The last observation leads us to consider a measure that evaluates the performance of a model by summing the areas delineated between the Qrecall curve of the examined model and the Qrecall curve of a random model (which is linear). Areas above the linear curve are added and areas below the linear curve are subtracted. The areas themselves are calculated by subtracting the Qrecall of a random classification from the Qrecall of the model's classification in every point as shown in Figure 7.6. The areas where the model performed better than a random guess increase the measure's value while the areas where the model performed worse than a random guess decrease it. If the last total computed area is divided in the area delineated between the optimum model Qrecall curve and the random model (linear) Qrecall curve, then it reaches the extent to which the potential is extracted, independently of the number of instances in the dataset.

Formally, the PEM measure is calculated as:

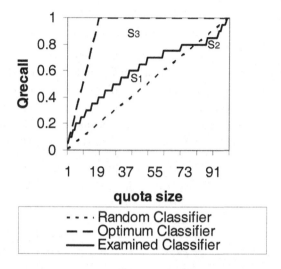

Fig. 7.6   A qualitative representation of PEM.

$$PEM = \frac{S_1 - S_2}{S_3} \qquad (7.20)$$

where S_1 is the area delimited by the Qrecall curve of the examined model above the Qrecall curve of a random model; S_2 is the area delimited by the Qrecall curve of the examined model under the Qrecall curve of a random model; and S_3 is the area delimited by the optimal Qrecall curve and the curve of the random model. The division in S_3 is required in order to normalize the measure, thus datasets of different size can be compared. In this way, if the model is optimal, then PEM gets the value 1. If the model is as good as a random choice, then the PEM gets the value 0. If it gives the worst possible result (that is to say, it puts the positive samples at the bottom of the list), then its PEM is -1. Based on the notations defined above, the PEM can be formulated as:

$$PEM = \frac{S_1 - S_2}{S_3} = \frac{\sum\limits_{j=1}^{n}\left(qrecall(j) - \frac{j}{n}\right)}{\sum\limits_{j=1}^{n^+}\left(\frac{j}{n^+}\right) + n^- - \sum\limits_{j=1}^{n}\left(\frac{j}{n}\right)} \qquad (7.21)$$

$$= \frac{\sum_{j=1}^{n} (qrecall(j)) - \frac{(n+1)}{2}}{\frac{(n^{+}+1)}{2} + n^{-} - \frac{(n+1)}{2}} = \frac{\sum_{j=1}^{n} (qrecall(j)) - \frac{(n+1)}{2}}{\frac{n^{-}}{2}} \qquad (7.22)$$

where n^{-} denotes the number of instances that are actually classified as "negative". Table 7.5 illustrates the calculation of PEM for the instances in Table 7.2. Note that the random Qrecall does not represent a certain realization but the expected values. The optimal qrecall is calculated as if the "positive" instances have been located in the top of the list.

Table 7.5   An Example for Calculating PEM for Instances of Table 7.2.

Place in list	Success probability	$t^{[k]}$	Model Qrecall	Random Qrecall	Optimal Qrecall	S1	S2	S3
1	0.45	1	0.25	0.1	0.25	0.15	0	0.15
2	0.34	0	0.25	0.2	0.5	0.05	0	0.3
3	0.32	1	0.5	0.3	0.75	0.2	0	0.45
4	0.26	1	0.75	0.4	1	0.35	0	0.6
5	0.15	0	0.75	0.5	1	0.25	0	0.5
6	0.14	0	0.75	0.6	1	0.15	0	0.4
7	0.09	1	1	0.7	1	0.3	0	0.3
8	0.07	0	1	0.8	1	0.2	0	0.2
9	0.06	0	1	0.9	1	0.1	0	0.1
10	0.03	0	1	1	1	0	0	0
					Total	1.75	0	3

Note that the PEM somewhat resembles the Gini index produced from Lorentz curves which appear in economics when dealing with the distribution of income. Indeed, this measure indicates the difference between the distribution of positive samples in a prediction and the uniform distribution. Note also that this measure gives an indication of the total lift of the model at every point. In every quota size, the difference between the Qrecall of the model and the Qrecall of a random model expresses the lift in extracting the potential of the test set due to the use in the model (for good or for bad).

7.1.7   *Statistical Tests for Comparing Ensembles*

Below we discuss some of the most common statistical methods proposed [Dieterich (1998)] for answering the following question: *Given two inducers A and B and a dataset S, which inducer will produce more accurate classifiers when trained on datasets of the same size?*

7.1.7.1 McNemar's Test

Let S be the available set of data, which is divided into a training set R and a test set T. Then we consider two inducers A and B trained on the training set and the result is two classifiers. These classifiers are tested on T and for each example $x \in T$ we record how it was classified. Thus, the contingency table presented in Table 7.6 is constructed.

Table 7.6 McNemar's Test: Contingency Table.

Number of examples misclassified by both classifiers (n_{00})	Number of examples misclassified by $\hat{f}_A$ but not by $\hat{f}_B(n_{01})$
Number of examples misclassified by $\hat{f}_B$ but not by $\hat{f}_A(n_{10})$	Number of examples misclassified neither by $\hat{f}_A$ nor by $\hat{f}_B(n_{11})$

The two inducers should have the same error rate under the null hypothesis H_0. McNemar's test is based on a χ^2 test for goodness-of-fit that compares the distribution of counts expected under null hypothesis to the observed counts. The expected counts under Ho are presented in Table 7.7.

Table 7.7 Expected Counts Under Ho.

n_{00}	$(n_{01} + n_{10})/2)$
$(n_{01} + n_{10})/2)$	$n_{11})$

The following statistic, s, is distributed as χ^2 with 1 degree of freedom. It incorporates a "continuity correction" term (of -1 in the numerator) to account for the fact that the statistic is discrete while the χ^2 distribution is continuous:

$$s = \frac{(|n_{10} - n_{01}| - 1)^2}{n_{10} + n_{01}} \tag{7.23}$$

According to the probabilistic theory [Athanasopoulos, 1991], if the null hypothesis is correct, the probability that the value of the statistic, s, is greater than $\chi^2_{1,0.95}$ is less than 0.05, i.e. $P(|s| > \chi^2_{1,0.95}) < 0.05$. Then, to compare the inducers A and B, the induced classifiers $\hat{f}_A$ and $\hat{f}_B$ are tested on T and the value of s is estimated as described above. Then if $|s| > \chi^2_{1,0.95}$, the null hypothesis could be rejected in favor of the hypothesis that the two inducers have different performance when trained on the particular training set R.

The shortcomings of this test are:

(1) It does not directly measure variability due to the choice of the training set or the internal randomness of the inducer. The inducers are compared using a single training set R. Thus McNemar's test should be only applied if we consider that the sources of variability are small.
(2) It compares the performance of the inducers on training sets, which are substantially smaller than the size of the whole dataset. Hence we must assume that the relative difference observed on training sets will still hold for training sets of size equal to the whole dataset.

7.1.7.2　*A Test for the Difference of Two Proportions*

This statistical test is based on measuring the difference between the error rates of algorithms A and B [Snedecor and Cochran (1989)]. More specifically, let $p_A = (n_{00} + n_{01})/n$ be the proportion of test examples incorrectly classified by algorithm A and let $p_B = (n_{00} + n_{10})/n$ be the proportion of test examples incorrectly classified by algorithm B. The assumption underlying this statistical test is that when algorithm A classifies an example x from the test set T, the probability of misclassification is p_A. Then the number of misclassifications of n test examples is a binomial random variable with mean np_A and variance $p_A(1 - p_A)n$.

The binomial distribution can be well approximated by a normal distribution for reasonable values of n. The difference between two independent normally distributed random variables is itself normally distributed. Thus, the quantity $p_A - p_B$ can be viewed as normally distributed if we assume that the measured error rates p_A and p_B are independent. Under the null hypothesis, Ho, the quantity $p_A - p_B$ has a mean of zero and a standard deviation error of

$$se = \sqrt{2p \cdot \left(1 - \frac{p_A + p_B}{2}\right)/n} \qquad (7.24)$$

where n is the number of test examples.

Based on the above analysis, we obtain the statistic:

$$z = \frac{p_A - p_B}{\sqrt{2p(1 - p)/n}} \qquad (7.25)$$

which has a standard normal distribution. According to the probabilistic theory, if the z value is greater than $Z_{0.975}$, the probability of incorrectly rejecting the null hypothesis is less than 0.05. Thus, if $|z| > Z_{0.975} = 1.96$, the null hypothesis could be rejected in favor of the hypothesis that the two algorithms have different performances. Two of the most important problems with this statistic are:

(1) The probabilities p_A and p_B are measured on the same test set and thus they are not independent.
(2) The test does not measure variation due to the choice of the training set or the internal variation of the learning algorithm. Also it measures the performance of the algorithms on training sets of a size significantly smaller than the whole dataset.

7.1.7.3 *The Resampled Paired t Test*

The resampled paired t test is the most popular in machine learning. Usually, there are a series of 30 trials in the test. In each trial, the available sample S is randomly divided into a training set R (it is typically two thirds of the data) and a test set T. The algorithms A and B are both trained on R and the resulting classifiers are tested on T. Let $p_A^{(i)}$ and $p_B^{(i)}$ be the observed proportions of test examples misclassified by algorithm A and B respectively during the i-th trial. If we assume that the 30 differences $p^{(i)} = p_A^{(i)} - p_B^{(i)}$ were drawn independently from a normal distribution, then we can apply Student's t test by computing the statistic:

$$t = \frac{\bar{p} \cdot \sqrt{n}}{\sqrt{\frac{\sum_{i=1}^{n} (p^{(i)} - \bar{p})^2}{n-1}}} \tag{7.26}$$

where $\bar{p} = \frac{1}{n} \cdot \sum_{i=1}^{n} p^{(i)}$. Under the null hypothesis, this statistic has a t distribution with $n - 1$ degrees of freedom. Then for 30 trials, the null hypothesis could be rejected if $|t| > t_{29,0.975} = 2.045$. The main drawbacks of this approach are:

(1) Since $p_A^{(i)}$ and $p_B^{(i)}$ are not independent, the difference $p^{(i)}$ will not have a normal distribution.
(2) The $p^{(i)}$'s are not independent, because the test and training sets in the trials overlap.

7.1.7.4   *The k-fold Cross-validated Paired t Test*

This approach is similar to the resampled paired t test except that instead of constructing each pair of training and test sets by randomly dividing S, the dataset is randomly divided into k disjoint sets of equal size, $T_1, T_2, \ldots, T_k$. Then k trials are conducted. In each trial, the test set is T_i and the training set is the union of all of the others T_j, $j \neq i$. The t statistic is computed as described in Section 7.1.7.3. The advantage of this approach is that each test set is independent of the others. However, there is the problem that the training sets overlap. This overlap may prevent this statistical test from obtaining a good estimation of the amount of variation that would be observed if each training set were completely independent of the others training sets.

7.2   Computational Complexity

Another useful criterion for comparing inducers and classifiers is their computational complexity. Strictly speaking computational complexity is the amount of CPU consumed by each inducer. It is convenient to differentiate between three metrics of computational complexity:

- Computational complexity for generating a new classifier: This is the most important metric, especially when there is a need to scale the data mining algorithm to massive datasets. Because most of the algorithms have computational complexity, which is worse than linear in the numbers of tuples, mining massive datasets might be prohibitively expensive.
- Computational complexity for updating a classifier: Given new data, what is the computational complexity required for updating the current classifier such that the new classifier reflects the new data?
- Computational complexity for classifying a new instance: Generally this type of metric is neglected because it is relatively small. However, in certain methods (like k-nearest neighborhood) or in certain real-time applications (like anti-missiles applications), this type can be critical.

A smaller ensemble requires less memory for storing its members. Moreover, smaller ensembles have a faster classification speed. It is particularly crucial in several near real-time applications, such as worm detection. In addition to the pursuing the highest possible accuracy, these applications, require that the classification time should be kept to the minimum.

7.3 Interpretability of the Resulting Ensemble

Interpretability (also known as comprehensibility) indicates the user ability to understand the ensemble results. While the generalization error measures how the classifier fits the data, comprehensibility measures the "mental fit" of that classifier. This is especially important in applications in which the user is required to understand the system behavior or explain its classification. For example the revised version of AdaBoost presented in [Friedman *et al.* (2000)] is considered to provide interpretable descriptions of the aggregate decision rule.

Interpretability is usually a subjective criterion. Nevertheless, there are several quantitative measures and indicators that can help us in evaluating this criterion. For example:

- Compactness - measures the knowledge representation size efficiency. Obviously, results presented by a smaller size are easier to understand. In ensemble methods compactness can be measured by the ensemble size (number of classifiers) and the complexity of each classifier. According to Freund and Mason [Freund and Mason (1999)] Even for modest values of ensemble size, boosting of decision trees could result in a final combined classifier with thousands (or millions) of nodes which it is difficult to visualize.
- Base Inducer Used - The base inducer used by the ensemble can determine its interpretability. Many techniques, like neural networks or support vector machines (SVM), are designed solely to achieve accuracy. However, as their classifiers are represented using large assemblages of real valued parameters, they are also difficult to understand and are referred to as black-box models. On the other hand, decision trees are easier to understand than black-box methods.

However it is often important for the researcher to be able to inspect an induced classifier. For such domains as medical diagnosis, users must understand how the system makes its decisions in order to be confident of the outcome. Since data mining can also play an important role in the process of scientific discovery, a system may discover salient features in the input data whose importance was not previously recognized. If the representations formed by the inducer are comprehensible, then these discoveries can be made accessible to human review [Hunter and Klein (1993)].

7.4   Scalability to Large Datasets

Scalability refers to the ability of the method to construct the classification model efficiently given large amounts of data. Classical induction algorithms have been applied with practical success in many relatively simple and small-scale problems. However, trying to discover knowledge in real life and large databases introduces time and memory problems.

There are ensemble methods (such as partitioning methods) that are more suitable to scale to large dataset than other. Moreover independent methods are considered to be more scalable than dependent methods because the former case, classifiers can be trained in parallel.

As large databases have become the norm in many fields (including astronomy, molecular biology, finance, marketing, health care, and many others), the use of data mining to discover patterns in them has become a potentially very productive enterprise. Many companies are staking a large part of their future on these "data mining" applications, and looking to the research community for solutions to the fundamental problems they encounter.

While a very large amount of available data used to be a dream of any data analyst, nowadays the synonym for "very large" has become "terabyte", a hardly imaginable volume of information. Information-intensive organizations (like telecom companies and banks) are supposed to accumulate several terabytes of raw data every one to two years.

However, the availability of an electronic data repository (in its enhanced form known as a "data warehouse") has caused a number of previously unknown problems, which, if ignored, may turn the task of efficient data mining into mission impossible. Managing and analyzing huge data warehouses requires special and very expensive hardware and software, which often causes a company to exploit only a small part of the stored data.

According to [Fayyad *et al.* (1996)] the explicit challenges for the data mining research community is to develop methods that facilitate the use of data mining algorithms for real-world databases. One of the characteristics of a real-world databases is high volume data.

Huge databases pose several challenges:

- Computing complexity: Since most induction algorithms have a computational complexity that is greater than linear in the number of attributes or tuples, the execution time needed to process such databases might become an important issue.

- Poor classification accuracy due to difficulties in finding the correct classifier. Large databases increase the size of the search space, and thus it increases the chance that the inducer will select an over fitted classifier that is not valid in general.

- Storage problems: In most machine learning algorithms, the entire training set should be read from the secondary storage (such as magnetic storage) into the computer's primary storage (main memory) before the induction process begins. This causes problems since the main memory's capability is much smaller than the capability of magnetic disks.

The difficulties in implementing classification algorithms as-is on high volume databases derives from the increase in the number of records/instances in the database and from the increase in the number of attributes/features in each instance (high dimensionality).

Approaches for dealing with a high number of records include:

- Sampling methods — statisticians are selecting records from a population by different sampling techniques.

- Aggregation — reduces the number of records either by treating a group of records as one, or by ignoring subsets of "unimportant" records.

- Massively parallel processing — exploiting parallel technology — to simultaneously solve various aspects of the problem.

- Efficient storage methods — enabling the algorithm to handle many records.

- Reducing the algorithm's search space.

7.5 Robustness

The ability of the model to handle noise or data with missing values and make correct predictions is called robustness. Different ensembles algorithms have different robustness levels. In order to estimate the robustness of an ensemble, it is common to train the ensemble on a clean training set and then train a different ensemble on a noisy training set. The noisy training set is usually the clean training set to which some artificial noisy instances have been added. The robustness level is measured as the difference in the accuracy of these two situations.

7.6 Stability

Formally, stability of a classification algorithm is defined as the degree to which an algorithm generates repeatable results, given different batches of data from the same process. In mathematical terms, stability is the expected agreement between two models on a random sample of the original data, where agreement on a specific example means that both models assign it to the same class. The instability problem raises questions about the validity of a particular ensemble provided as an output of a given algorithm. The users view the learning algorithm as an oracle. Obviously, it is difficult to trust an oracle that says something radically different each time you make a slight change in the data.

7.7 Flexibility

Flexibility indicates the ability to use any inducer (inducer-independent), any combiner (Combiner-independent), provide a solution to variety of classification tasks (for example it is should not be limited to a binary classification task), a set of controlling parameters which enable the user to examine several variations of the ensemble techniques.

7.8 Usability

Machine learning is highly iterative process. Practitioners typically adjust algorithm's parameters to generate better classifiers. A good ensemble method should provide a set of controlling parameters that are comprehensive and can be easily tuned.

7.9 Software Availability

Software Availability of an ensemble method indicates how many off-the-shelf software packages support this ensemble method. High Availability implies that the practitioner can move from one software to another, without the need to replace his ensemble method. Table 7.8 indicates the popularity (as measured by the number of citation in Google scholar in June, 2009) of the ten methods presented in Table 7.8 and the total availability of these methods in five open source packages: Weka [Frank et. al

(2005)], Orange [Demsar *et al.* (2004)], Tanagra [Rakotomalala (2005)], RapidMiner (formerly YALE)[Mierswa *et al.* (2006)], OpenDT [Banfield (2005)], Java-ML /citeAbeel and R programming environment [R statistical computing Language (2005)]. The table indicates that high popularity is a necessary condition for high availability, but still there are popular methods with relatively low availability.

In addition, there are several open source packages which specifically implement variants of boosting algorithms. OAIDTB (Other Application Interactively Demonstrating Techniques of Boosting)[Villalba *et al.* (2003)] extends Weka framework by adding the following implementations: AdaBoost.real, AdaBoost.M1W, GentleAdaBoost, AdaBoost.OC, AdaCost, AdaBoost.ECC, CSBx, AdaBoost.MH and CSAdaBoostMH. The package JBoost (http://jboost.sourceforge.net/) implements algorithms such as AdaBoost, LogitBoost, RobustBoost, Boostexter and BrownBoost (included in JBoost 1.4). A source code of Boostexter can be also obtained from: http://www.cs.princeton.edu/ schapire/boostexter.html.

Table 7.8   Software Availability of the Ensemble Method.

Algorithm	Google Scholar (June 2009)	Software Availability	Reference
AdaBoost	2730	Weka, Orange, Tanagra, RapidMiner, R	[Freund and Schapire (1996)]
Bagging	4907	All	[Breiman (1996a)]
RandomForest	1988	All	[Breiman (2001)]
DECORATE	81	Weka	[Melville and Mooney (2003)]
MultiBoosting	184	Weka, RapidMiner	[Webb (2000)]
Wagging	993	Weka, RapidMiner	[Bauer and Kohavi (1999)]
Attribute Bagging	52	Weka	[Bryll *et al.* (2003)]
Stacking	1582	Weka, Tanagra, RapidMiner	[Wolpert (1992)]
ECOC	1007	Weka, RapidMiner	[Dietterich and Bakiri (1995)]
Arc-x4	644	Weka, Tanagra	[Breiman (1998)]

7.10   Which Ensemble Method Should be Used?

Given the vast repertoire of ensemble methods to choose from, and the various potentially contradicting criteria, it is not surprising that choosing an ensemble method is not a simple task.

Empirical comparison of the performance of different approaches and their variants in a wide range of application domains has shown that each performs best in some, but not all, domains. This has been termed the selective superiority problem [Brodley (1995a)].

It is well known that no induction algorithm can be the best in all possible domains; each algorithm contains an explicit or implicit bias [Mitchell (1980)] that leads it to prefer certain generalizations over others. The algorithm will be successful only insofar as this bias matches the characteristics of the application domain [Brazdil *et al.* (1994)]. Furthermore, other results have demonstrated the existence and correctness of the "conservation law" [Schaffer (1994)] or "no free lunch theorem" [Wolpert (1996)]: if one inducer is better than another in some domains, then there are necessarily other domains in which this relationship is reversed.

The "no free lunch theorem" implies that for a given problem, a certain approach can yield more information from the same data than other approaches.

A distinction should be made between all the mathematically possible domains, which are simply a product of the representation languages used, and the domains that occur in the real world, and are therefore the ones of primary interest [Rao *et al.* (1995)]. Without doubt there are many domains in the former set that are not in the latter, and average accuracy in the realworld domains can be increased at the expense of accuracy in the domains that never occur in practice. Indeed, achieving this is the goal of inductive learning research. It is still true that some algorithms will match certain classes of naturallyoccurring domains better than other algorithms, and so achieve higher accuracy than these algorithms. While this may be reversed in other realworld domains, it does not preclude an improved algorithm from being as accurate as the best in each of the domain classes.

Indeed, in many application domains, the generalization error of even the best methods is far above 0%, and the question of whether it can be improved, and if so how, is an open and important one. One aspect in answering this question is determining the minimum error achievable by any classifier in the application domain (known as the optimal Bayes error). If existing classifiers do not reach this level, new approaches are needed. Although this problem has received considerable attention (see for instance [Tumer and Ghosh (1996)]), no generally reliable method has so far been demonstrated.

The "no free lunch" concept presents a dilemma to the analyst approaching a new task: Which inducer should be used?

If the analyst is looking for accuracy only, one solution is to try each one in turn, and by estimating the generalization error, to choose the one that appears to perform best [Schaffer (1994)]. Another approach, known as *multistrategy learning* [Michalski and Tecuci (1994)], attempts to combine two or more different paradigms in a single algorithm. Most research in this area has been concerned with combining empirical approaches with analytical methods (see for instance [Towell and Shavlik (1994)]. Ideally, a multistrategy learning algorithm would always perform as well as the best of its "parents" obviating the need to try each one and simplifying the knowledge acquisition task. Even more ambitiously, there is hope that this combination of paradigms might produce synergistic effects (for instance by allowing different types of frontiers between classes in different regions of the example space), leading to levels of accuracy that neither atomic approach by itself would be able to achieve.

Unfortunately, this approach has often been used with only moderate success. Although it is true that in some industrial applications (like in the case of demand planning) this strategy proved to boost the error performance, in many other cases the resulting algorithms are prone to be cumbersome, and often achieve an error that lies between those of their parents, instead of matching the lowest. The dilemma of what method to choose becomes even greater, if many criteria are taken into consideration.

The difficulty in choosing the ensemble methods results from the fact that this is a MCDM (Multiple Criteria Decision Making) problem. There are trade off relationships among the criteria and some criteria can not be measured in commensurate units. Thus, in order to systematically chose the right method, the practitioner is encouraged to implement one of the MCDM solving technique such as AHP (Analytic Hierarchy Process).

Moreover, the context of the specific classification problem to be solved has tremendous effect on the results. In general, all comparative studies that have been performed in the literature and aim to compare the predictive performance, show that the no-free-lunch theorem holds [Brown *et al.* (2005); Sohna (2007)], i.e. the best ensemble technique depends much on the particular training dataset. Thus, the current challenge is to automatically choose the best ensemble technique. There are two alternatives to achieve this goal:

- The wrapper approach – Given a certain dataset, use each ensemble method and select the one that appears to give the highest success rate. The main advantage of this approach is its ability to predict quite well the performance of each examined method. The main disadvantage of this method is it's prolonged processing time. For some inducers the induction times may be very long, particularly in large real-life datasets. Several researchers have implemented this approach for selecting induction algorithms or dimension reduction algorithms and showed that it produces superior results [Schaffer (1993)].

- The meta-learning approach [Vilalta *et al.* (2005)]– Based on datasets characteristics, the meta-classifier decides whether to use ensemble method or not and what technique to use. If a certain ensemble method outperforms other methods in a particular dataset, then one should expect that this method will be preferable when other problems with similar characteristics are presented. For this purpose one can employ meta-learning. Meta-learning is concerned with accumulating experience on the performance of multiple applications of a learning system. One possible output of the meta-learning process is a meta-classifier that is capable to indicate which learning method is most appropriate to a given problem. This goal can be accomplished by performing the following phases: In the first phase one should examine the performance of all investigated ensemble methods on various datasets. Upon examination of each dataset, the characteristics of the dataset are extracted. The dataset's characteristics, together with the indication of the most preferable ensemble method, (in this dataset) are stored in a meta-dataset. This meta-dataset reflects the experience accumulated across different datasets. In the second phase, an inducer can be applied to this meta-dataset to induce a meta-classifier that can map a dataset to the most appropriate ensemble method (based on the characteristics of the dataset). In the last phase, the meta-classifier is actually used to match a new unseen dataset to the most appropriate ensemble method. Several researchers have implemented this approach for selecting an ensemble method and showed that it produces superior results [Rokach (2006)]

Bibliography

Abeel Thomas, Yves Van de Peer, Yvan Saeys, Java-ML: A Machine Learning Library, Journal of Machine Learning Research 10 (2009) 931-934

Adem, J., Gochet, W., 2004. Aggregating classifiers with mathematical programming. Comput. Statist. Data Anal. 47 (4), 791-807.

Aha, D. W.; Kibler, D.; and Albert, M. K., Instancebased learning algorithms. Machine Learning 6(1):37-66, 1991.

Ahn H., Moon H., Fazzari M. J., Noha Lim,James J. Chen, Ralph L. Kodell, Classification by ensembles from random partitions of high-dimensional data, Computational Statistics and Data Analysis 51 (2007) 6166-6179

Al-Sultan K. S. , Khan M. M. : Computational experience on four algorithms for the hard clustering problem. Pattern Recognition Letters 17(3): 295-308, 1996.

Al-Sultan K. S., A tabu search approach to the clustering problem, Pattern Recognition, 28:1443-1451,1995.

Alba, E., Chicano, J.F., (2004), Solving the error correcting code problem with parallel hybrid heuristics. In: Proceedings of 2004 ACM Symposium on Applied Computing. Volume 2. 985–989.

Alba, E., Cotta, C., Chicano, F., Nebro, A.J., (2002), Parallel evolutionary algorithms in telecommunications: two case studies. In: Proceedings of Congresso Argentino de Ciências de la Computación.

Ali K. M., Pazzani M. J., Error Reduction through Learning Multiple Descriptions, Machine Learning, 24: 3, 173-202, 1996.

Allwein, E.L., Shapire, R.E., Singer, Y., (2000), Reducing multiclass to binary: a unifying approach for magin classifiers. In: Proceedings of the 17th International Conference on Machine Learning, Morgan Kaufmann 9–16.

Almuallim H,. and Dietterich T.G., Learning Boolean concepts in the presence of many irrelevant features. Artificial Intelligence, 69: 1-2, 279-306, 1994.

Almuallim H., An Efficient Algorithm for Optimal Pruning of Decision Trees. Artificial Intelligence 83(2): 347-362, 1996.

Alpaydin, E., Mayoraz, E., (1999), Learning error-correcting output codes from data. In: Proceedings of the 9th International Conference on Neural Networks. 743–748.

Alsabti K., Ranka S. and Singh V., CLOUDS: A Decision Tree Classifier for Large
Datasets, Conference on Knowledge Discovery and Data Mining (KDD-98),
August 1998.

Altincay H., Decision trees using model ensemble-based nodesPattern Recognition
40 (2007) 3540 - 3551.

An A. and Wang Y., "Comparisons of classification methods for screening potential compounds". *In IEEE International Conference on Data Mining,
2001.*

Anand R, Methrotra K, Mohan CK, Ranka S. Efficient classification for multiclass
problems using modular neural networks. IEEE Trans Neural Networks,
6(1): 117-125, 1995.

Anderson, J.A. and Rosenfeld, E. Talking Nets: An Oral History of Neural Network Research. Cambridge, MA: MIT Press, 2000.

Arbel, R. and Rokach, L., Classifier evaluation under limited resources, Pattern
Recognition Letters, 27(14): 1619–1631, 2006, Elsevier.

Archer K. J., Kimes R. V., Empirical characterization of random forest variable importance measures, Computational Statistics and Data Analysis 52
(2008) 2249-2260

Ashenhurst, R. L., The decomposition of switching functions, Technical report,
Bell Laboratories BL-1(11), pp. 541-602, 1952.

Athanasopoulos, D. (1991). *Probabilistic Theory.* Stamoulis, Piraeus.

Attneave F., Applications of Information Theory to Psychology. Holt, Rinehart
and Winston, 1959.

Averbuch, M. and Karson, T. and Ben-Ami, B. and Maimon, O. and Rokach, L.,
Context-sensitive medical information retrieval, The 11th World Congress
on Medical Informatics (MEDINFO 2004), San Francisco, CA, September
2004, IOS Press, pp. 282-286

Averbuch M., Maimon O., Rokach L., and Ezer E., Free-Text Information Retrieval System for a Rapid Enrollment of Patients into Clinical Trials, Clinical Pharmacology and Therapeutics, 77(2): 13-14, 2005.

Avnimelech R. and Intrator N., Boosted Mixture of Experts: an ensemble learning
scheme, Neural Computations, 11(2):483-497, 1999.

Bäck, T., Fogel, D.B., Michalewicz, T., (2000), Evolutionary Computation 1:
Basic Algorithms and Operators. Institute of Physics Publishing.

Baker E., and Jain A. K., On feature ordering in practice and some finite sample effects. In Proceedings of the Third International Joint Conference on
Pattern Recognition, pages 45-49, San Diego, CA, 1976.

Bala J., Huang J., Vafaie H., De Jong K., Wechsler H., Hybrid Learning Using
Genetic Algorithms and Decision Trees for Pattern Classification, IJCAI
conference, 1995.

Banfield R., OpenDT, http://opendt.sourceforge.net/, 2005.

Banfield J. D. and Raftery A. E. . Model-based Gaussian and non-Gaussian clustering. Biometrics, 49:803-821, 1993.

Robert E. Banfield, Lawrence O. Hall, Kevin W. Bowyer, W.P. Kegelmeyer, A
Comparison of Decision Tree Ensemble Creation Techniques, IEEE Transactions on Pattern Analysis and Machine Intelligence, vol. 29, no. 1, pp.

173-180, Jan., 2007

Bao Y., Ishii N., Combining multiple K-nearest neighbor classifiers for text classification by reducts. In: Proceedings of 5th international conference on discovery science, LNCS 2534, Springer, 2002, pp 340-347

Bartlett P. and Shawe-Taylor J., Generalization Performance of Support Vector Machines and Other Pattern Classifiers, In "Advances in Kernel Methods, Support Vector Learning", Bernhard Scholkopf, Christopher J. C. Burges, and Alexander J. Smola (eds.), MIT Press, Cambridge, USA, 1998.

Bartlett, P., Traskin, M., 2007. Adaboost is consistent. Journal of Machine Learning Research 8, 2347-2368.

Basak J., Online adaptive decision trees, Neural Computations, 16(9):1959-1981, 2004.

Basak J., Online Adaptive Decision Trees: Pattern Classification and Function Approximation, Neural Computations, 18(9):2062-2101, 2006.

Bauer, E. and Kohavi, R., "An Empirical Comparison of Voting Classification Algorithms: Bagging, Boosting, and Variants". Machine Learning, 35: 1-38, 1999.

Baxt, W. G., Use of an artificial neural network for data analysis in clinical decision making: The diagnosis of acute coronary occlusion. Neural Computation, 2(4):480-489, 1990.

Bay, S., Nearest neighbor classification from multiple feature subsets. Intelligent Data Analysis, 3(3): 191-209, 1999.

Beasley, D. (2000), [Bäck *et al.* (2000)] 4–18

Bellman, R., Adaptive Control Processes: A Guided Tour, Princeton University Press, 1961.

Bennett X. and Mangasarian O.L., Multicategory discrimination via linear programming. Optimization Methods and Software, 3:29-39, 1994.

Kristin P. Bennett and Ayhan Demiriz and Richard Maclin, Exploiting unlabeled data in ensemble methods, Proceedings of the eighth ACM SIGKDD international conference on Knowledge discovery and data mining, pp. 289–296, ACM Press, New York, NY, USA, 2002.

Bensusan H. and Kalousis A., Estimating the Predictive Accuracy of a Classifier, In Proc. Proceedings of the 12th European Conference on Machine Learning, pages 25-36, 2001.

Bentley J. L. and Friedman J. H., Fast algorithms for constructing minimal spanning trees in coordinate spaces. IEEE Transactions on Computers, C-27(2):97-105, February 1978. 275

BenBassat M., Myopic policies in sequential classification. IEEE Trans. on Computing, 27(2):170-174, February 1978.

Berger, A., (1999), Error-correcting output coding for text classification.

Bernard M.E., Decision trees and diagrams. Computing Surveys, 14(4):593-623, 1982.

Berry M., and Linoff G., Mastering Data Mining, John Wiley & Sons, 2000.

Bhargava H. K., Data Mining by Decomposition: Adaptive Search for Hypothesis Generation, INFORMS Journal on Computing Vol. 11, Iss. 3, pp. 239-47, 1999.

Biermann, A. W., Faireld, J., and Beres, T. (1982). Signature table systems and learning. IEEE Trans. Syst. Man Cybern., 12(5):635-648.

Black, M. and Hickey, R.J., Maintaining the Performance of a Learned Classifier under Concept Drift, Intelligent Data Analysis 3(1),pp. 453474, 1999.

Blake, C.L., Merz, C.J., (1998), UCI repository of machine learning databases.

Blum, A. L. and Langley, P., 1997, Selection of relevant features and examples in machine learning, Artificial Intelligence, 97, pp.245-271.

Blum A., and Mitchell T., Combining Labeled and Unlabeled Data with CoTraining. In Proc. of the 11th Annual Conference on Computational Learning Theory, pages 92-100, 1998.

Bonner, R., On Some Clustering Techniques. IBM journal of research and development, 8:22-32, 1964.

Booker L., Goldberg D. E., and Holland J. H., Classifier systems and genetic algorithms. Artificial Intelligence, 40(1-3):235-282, 1989.

Boser, R.C., Ray-Chaudhuri, D.K., (1960), On a class of error-correcting binary group codes. Information and Control 3 68–79.

Brachman, R. and Anand, T., 1994, The process of knowledge discovery in databases, in: Advances in Knowledge Discovery and Data Mining, AAAI/MIT Press, pp. 37-58.

Bratko I., and Bohanec M., Trading accuracy for simplicity in decision trees, Machine Learning 15: 223-250, 1994.

Brazdil P., Gama J., Henery R., Characterizing the Applicability of Classification Algorithms using Meta Level Learning, in Machine Learning: ECML-94, F.Bergadano e L. de Raedt (eds.), LNAI No. 784: pp. 83-102, Springer-Verlag, 1994.

Breiman, L., Random forests. Machine Learn-ing, 45, 532, 2001.

Breiman L., Friedman J., Olshen R., and Stone C.. Classification and Regression Trees. Wadsworth Int. Group, 1984.

Breiman L. (1996a), Bagging predictors, Machine Learning, 24(2):123-140, 1996.

Breiman L. (1996b), Stacked regressions, Machine Learning, 24(2):4964, 1996.

Breiman L., Arcing classifiers, Annals of Statistics, vol. 26,no. 3, pp. 801-849, 1998.

Breiman, L., Pasting small votes for classification in large databases and on-line.Machine Learning, 36, 85-103.

Breiman, L., Randomizing outputs to increase prediction accuracy, Mach. Learn. 40 (3) (2000) 229-242.

Brodley C. E. and Utgoff. P. E., Multivariate decision trees. Machine Learning, 19:45-77, 1995.

Brodley, C. E., Automatic selection of split criterion during tree growing based on node selection. In Proceedings of the Twelth International Conference on Machine Learning, 73-80 Taho City, Ca. Morgan Kaufmann, 1995a.

Brodley C.E., Recursive automatic bias selection for classifier construction, Machine Learning 20 (1995b) 63-94.

Brown G., Wyatt J. L., Negative Correlation Learning and the Ambiguity Family of Ensemble Methods. Multiple Classifier Systems 2003: 266–275

Brown G., Wyatt J., Harris R., Yao X., Diversity creation methods: a survey and categorisation, Information Fusion, 6(1):5–20.

Bruzzone L., Cossu R., Vernazza G., Detection of land-cover transitions by combining multidate classifiers, Pattern Recognition Letters, 25(13): 1491–1500, 2004.

R. Bryll, Gutierrez-Osuna R., Quek F., Attribute bagging: improving accuracy of classifier ensembles by using random feature subsets, Pattern Recognition Volume 36 (2003): 1291-1302

Buchanan, B.G. and Shortliffe, E.H., Rule Based Expert Systems, 272-292, Addison-Wesley, 1984.

Buczak A. L. and Ziarko W., "Neural and Rough Set Based Data Mining Methods in Engineering", Klosgen W. and Zytkow J. M. (Eds.), Handbook of Data Mining and Knowledge Discovery, pages 788-797. Oxford University Press, 2002.

Buja, A. and Lee, Y.S., Data mining criteria for tree based regression and classification, Proceedings of the 7th International Conference on Knowledge Discovery and Data Mining, (pp 27-36), San Diego, USA, 2001.

Buntine W., Niblett T., A Further Comparison of Splitting Rules for Decision-Tree Induction. Machine Learning, 8: 75-85, 1992.

Buntine, W., A Theory of Learning Classification Rules. Doctoral dissertation. School of Computing Science, University of Technology. Sydney. Australia, 1990.

Buntine, W. (1992), "Learning Classification Trees", Statistics and Computing, 2, 63–73.

Buntine, W., "Graphical Models for Discovering Knowledge", in U. Fayyad, G. Piatetsky-Shapiro, P. Smyth, and R. Uthurusamy, editors, Advances in Knowledge Discovery and Data Mining, pp 59-82. AAAI/MIT Press, 1996.

Buttrey, S.E., Karo, C., 2002. Using k-nearest-neighbor classification in the leaves of a tree. Comput. Statist. Data Anal. 40, 27-37.

Buhlmann, P. and Yu, B., Boosting with L_2 loss: Regression and classification, Journal of the American Statistical Association, 98, 324338. 2003.

Califf M.E. and Mooney R.J., Relational learning of pattern-match rules for information extraction. Proceedings of the Sixteenth National Conf. on Artificial Intelligence, page 328-334, 1999.

Can F. , Incremental clustering for dynamic information processing, in ACM Transactions on Information Systems, no. 11, pp 143-164, 1993.

Cantu-Paz E., Kamath C., Inducing oblique decision trees with evolutionary algorithms, IEEE Trans. on Evol. Computation 7(1), pp. 54-68, 2003.

Cardie, C. (1995). Using decision trees to improve cased- based learning. In *Proceedings of the First International Conference on Knowledge Discovery and Data Mining*. AAAI Press.

Caropreso M., Matwin S., and Sebastiani F., A learner-independent evaluation of the useful-ness of statistical phrases for automated text categorization, Text Databases and Document Management: Theory and Practice. Idea Group Publishing , page 78-102, 2001.

Caruana R., Niculescu-Mizil A. , Crew G. , Ksikes A., Ensemble selection from libraries of models, Twenty-first international conference on Machine learning, July 04-08, 2004, Banff, Alberta, Canada.

Carvalho D.R., Freitas A.A., A hybrid decision-tree - genetic algorithm method for data mining, Information Science 163, 13-35, 2004.

Catlett J., Mega induction: Machine Learning on Vary Large Databases, PhD, University of Sydney, 1991.

Chai B., Huang T., Zhuang X., Zhao Y., Sklansky J., Piecewise-linear classifiers using binary tree structure and genetic algorithm, Pattern Recognition 29(11), pp. 1905-1917, 1996.

Chan P. K. and Stolfo, S. J., Toward parallel and distributed learning by metalearning, In AAAI Workshop in Knowledge Discovery in Databases, pp. 227-240, 1993.

Chan P.K. and Stolfo, S.J., A Comparative Evaluation of Voting and Metalearning on Partitioned Data, Proc. 12th Intl. Conf. On Machine Learning ICML-95, 1995.

Chan P.K. and Stolfo S.J, On the Accuracy of Meta-learning for Scalable Data Mining, J. Intelligent Information Systems, 8:5-28, 1997.

Charnes, A., Cooper, W. W., and Rhodes, E., Measuring the efficiency of decision making units, European Journal of Operational Research, 2(6):429-444, 1978.

Chawla N. V., Moore T. E., Hall L. O., Bowyer K. W., Springer C., and Kegelmeyer W. P.. Distributed learning with bagging-like performance. Pattern Recognition Letters, 24(1-3):455-471, 2002.

Chawla N. V., Hall L. O., Bowyer K. W., Kegelmeyer W. P., Learning Ensembles from Bites: A Scalable and Accurate Approach, The Journal of Machine Learning Research archive, 5:421–451, 2004.

Cheeseman P., Stutz J.: Bayesian Classification (AutoClass): Theory and Results. Advances in Knowledge Discovery and Data Mining 1996: 153-180

Chen K., Wang L. and Chi H., Methods of Combining Multiple Classifiers with Different Features and Their Applications to Text-Independent Speaker Identification, International Journal of Pattern Recognition and Artificial Intelligence, 11(3): 417-445, 1997.

Cherkauer, K. J. and Shavlik, J. W., Growing simpler decision trees to facilitate knowledge discovery. In *Proceedings of the Second International Conference on Knowledge Discovery and Data Mining.* AAAI Press, 1996.

Cherkauer, K.J., Human Expert-Level Performance on a Scientific Image Analysis Task by a System Using Combined Artificial Neural Networks. In Working Notes, Integrating Multiple Learned Models for Improving and Scaling Machine Learning Algorithms Workshop, Thirteenth National Conference on Artificial Intelligence. Portland, OR: AAAI Press, 1996.

Chizi, B., Maimon, O. and Smilovici A. On Dimensionality Reduction of High Dimensional Data Sets, Frontiers in Artificial Intelligence and Applications, IOS press, pp. 230-236, 2002.

Christensen S. W. , Sinclair I., Reed P. A. S., Designing committees of models through deliberate weighting of data points, The Journal of Machine Learning Research, 4(1):39–66, 2004.

Christmann A., Steinwart I., Hubert M.,Robust learning from bites for data mining, Computational Statistics and Data Analysis 52 (2007) 347-361

Cios K. J. and Sztandera L. M., Continuous ID3 algorithm with fuzzy entropy measures, Proc. IEEE Internat. Con/i on Fuzz)' Systems,1992, pp. 469-476.

Clark, P. and Boswell, R., "Rule induction with CN2: Some recent improvements." In Proceedings of the European Working Session on Learning, pp. 151-163, Pitman, 1991.

Clark P., and Niblett T., The CN2 rule induction algorithm. Machine Learning, 3:261-284, 1989.

Clearwater, S., T. Cheng, H. Hirsh, and B. Buchanan. Incremental batch learning. In Proceedings of the Sixth International Workshop on Machine Learning, San Mateo CA:, pp. 366-370. Morgan Kaufmann, 1989.

Clemen R., Combining forecasts: A review and annotated bibliography. International Journal of Forecasting, 5:559–583, 1989

Cohen S., Rokach L., Maimon O., Decision Tree Instance Space Decomposition with Grouped Gain-Ratio, Information Science, Volume 177, Issue 17, pp. 3592-3612, 2007.

Collins, M., Shapire, R.E., Singer, Y., (2002), Logistic regression, adaboost and bregman distances. Machine Learning **47**(2/3) 253–285.

Coppock D. S., Data Modeling and Mining: Why Lift?, Published in DM Review online, June 2002.

Crammer, K., Singer, Y., (2002), On the learnability and design of output codes for multiclass problems. Machine Learning **47**(2-3) 201–233.

Crawford S. L., Extensions to the CART algorithm. Int. J. of ManMachine Studies, 31(2):197-217, August 1989.

Cristianini, N., Shawe-Taylor, J., (2000), An introduction to Support Vector Machines and other kernel-based learning methods. Cambridge University Press.

Croux C., Joossens K., Lemmens A., Trimmed bagging, Computational Statistics and Data Analysis 52 (2007) 362-368

Cunningham P., and Carney J., Diversity Versus Quality in Classification Ensembles Based on Feature Selection, In: R. L. de Mntaras and E. Plaza (eds.), Proc. ECML 2000, 11th European Conf. On Machine Learning,Barcelona, Spain, LNCS 1810, Springer, 2000, pp. 109-116.

Curtis, H. A., A New Approach to the Design of Switching Functions, Van Nostrand, Princeton, 1962.

Cutzu F., Polychotomous classification with pairwise classifiers: A new voting principle. In Proc. 4th International Workshop on Multiple Classifier Systems (MCS 2003), Lecture Notes in Computer Science, Guildford, UK, 2003, Vol. 2709, pp. 115-124.

Dasarathy B.V. and Sheela B.V., Composite classifier system design: Concepts and methodology, Proceedings of the IEEE, vol. 67, no. 5, pp. 708-713, 1979.

Džeroski S., Ženko B., Is Combining Classifiers with Stacking Better than Selecting the Best One?, Machine Learning, 54(3): 255–273, 2004.

Darwin, C., (1859), On the origin of species by means of natural selection. John Murray, London.

Deb, K., (2000), An efficient constraint handling method for genetic algorithms. Computer Methods in Applied Mechanics and Engineering **186** 311–338.

Dekel, O., Singer, Y., (2003), Multiclass learning by probabilistic embeddings. In: Advances in Neural Information Processing Systems. Volume 15., MIT Press 945–952.

Dempster A.P., Laird N.M., and Rubin D.B., Maximum likelihood from incomplete data using the EM algorithm. Journal of the Royal Statistical Society, 39(B), 1977.

Demsar J., Zupan B., Leban G. (2004) Orange: From ExperimentalMachine Learning to Interactive Data Mining, White Paper(www.ailab.si/orange), Faculty of Computer and InformationScience, University of Ljubljana.

Denison D.G.T., Adams N.M., Holmes C.C., Hand D.J., Bayesian partition modelling, Computational Statistics and Data Analysis 38 (2002) 475-485

Derbeko P. , El-Yaniv R. and Meir R., Variance optimized bagging, European Conference on Machine Learning, 2002.

Dhillon I. and Modha D., Concept Decomposition for Large Sparse Text Data Using Clustering. Machine Learning. 42, pp.143-175. (2001).

Dietterich, T. G., and Ghulum Bakiri. Solving multiclass learning problems via error-correcting output codes. Journal of Artificial Intelligence Research, 2:263-286, 1995.

Dietterich, T. G., and Kong, E. B., Machine learning bias, statistical bias, and statistical variance of decision tree algorithms. Tech. rep., Oregon State University, 1995.

Dietterich, T. G., and Michalski, R. S., A comparative review of selected methods for learning from examples, Machine Learning, an Artificial Intelligence approach, 1: 41-81, 1983.

Dietterich, T. G., Kearns, M., and Mansour, Y., Applying the weak learning framework to understand and improve C4.5. Proceedings of the Thirteenth International Conference on Machine Learning, pp. 96-104, San Francisco: Morgan Kaufmann, 1996.

Dietterich, T. G., "Approximate statistical tests for comparing supervised classification learning algorithms". Neural Computation, 10(7): 1895-1924, 1998.

Dietterich, T. G., An Experimental Comparison of Three Methods for Constructing Ensembles of Decision Trees: Bagging, Boosting and Randomization. 40(2):139-157, 2000.

Dietterich T., Ensemble methods in machine learning. In J. Kittler and F. Roll, editors, First International Workshop on Multiple Classifier Systems, Lecture Notes in Computer Science, pages 1-15. Springer-Verlag, 2000

Dimitrakakis C., Bengio S., Online adaptive policies for ensemble classifiers,Neurocomputing 64:211-221, 2005.

Dimitriadou E., Weingessel A., Hornik K., A cluster ensembles framework, Design and application of hybrid intelligent systems, IOS Press, Amsterdam, The Netherlands, 2003.

Domingos, P. and Hulten, G., Mining Time-Changing Data Streams, Proc. of KDD-2001, ACM Press, 2001.

Domingos, P., & Pazzani, M., On the Optimality of the Naive Bayes Classifier under Zero-One Loss, Machine Learning, 29: 2, 103-130, 1997.

Domingos, P., Using Partitioning to Speed Up Specific-to-General Rule Induction. In Proceedings of the AAAI-96 Workshop on Integrating Multiple Learned Models, pp. 29-34, AAAI Press, 1996.

Dominigos P. (1999): *MetaCost: A general method for making classifiers cost sensitive*. In proceedings of the Fifth International Conference on Knowledge Discovery and Data Mining, pp. 155-164. ACM Press.

Domingo, C., and Watanabe, O. (2000). Madaboost: A modification of adaboost. colt2000 pp. 180-189.

Dontas, K., Jong, K.D., (1990), Discovery of maximal distance codes using genetic algorithms. In: Proceedings of the 2nd International IEEE Conference on Tools for Artificial Intelligence, IEEE Computer Society Press 905-811.

Dougherty, J., Kohavi, R, Sahami, M., Supervised and unsupervised discretization of continuous attributes. Machine Learning: Proceedings of the twelfth International Conference, Morgan Kaufman pp. 194-202, 1995.

Drucker H., Effect of pruning and early stopping on performance of a boosting ensemble, Computational Statistics and Data Analysis 38 (2002) 393-406

Duda, R., and Hart, P., Pattern Classification and Scene Analysis, New-York, Wiley, 1973.

Duda, P. E. Hart and D. G. Stork, Pattern Classification, Wiley, New York, 2001.

Duin, R. P. W., The combining classifier: to train or not to train? In Proc. 16th International Conference on Pattern Recognition, ICPR02, Canada, 2002, pp. 765-770.

Dunteman, G.H., Principal Components Analysis, Sage Publications, 1989.

Eiben, A.E., Smith, J.E., (2003), Introduction to Evolutionary Computing. Springer.

Elder I. and Pregibon, D., "A Statistical Perspective on Knowledge Discovery in Databases", In U. Fayyad, G. Piatetsky-Shapiro, P. Smyth, and R. Uthurusamy editors., Advances in Knowledge Discovery and Data Mining, pp. 83-113, AAAI/MIT Press, 1996.

Escalera, S., Pujol, O., Radeva, R., (2006), Decoding of ternary error correcting output codes. In: Proceedings of the 11th Iberoamerican Congress on Pattern Recognition. Volume 4225 of Lecture Notes in Computer Science., Springer-Verlag 753-763.

Esmeir, S., and Markovitch, S. 2004. Lookahead-basedalgorithms for anytime induction of decision trees. InICML04, 257264.

Esposito F., Malerba D. and Semeraro G., A Comparative Analysis of Methods for Pruning Decision Trees. EEE Transactions on Pattern Analysis and Machine Intelligence, 19(5):476-492, 1997.

Ester M., Kriegel H.P., Sander S., and Xu X., A density-based algorithm for discovering clusters in large spatial databases with noise. In E. Simoudis, J. Han, and U. Fayyad, editors, Proceedings of the 2nd International Conference on Knowledge Discovery and Data Mining (KDD-96), pages 226-231, Menlo Park, CA, 1996. AAAI, AAAI Press.

Estivill-Castro, V. and Yang, J. A Fast and robust general purpose clustering algorithm. Pacific Rim International Conference on Artificial Intelligence, pp. 208-218, 2000.

Fürnkranz, J., (2002), Round robin classification. Journal of Machine Learning Research **2** 721–747.

Fan, W. and Stolfo, S.J. and Zhang, J. and Chan, P.K. (1999), AdaCost: Misclassication Cost-sensitive Boosting, ICML 1999, pp. 97-105.

Fraley C. and Raftery A.E., "How Many Clusters? Which Clustering Method? Answers Via Model-Based Cluster Analysis", Technical Report No. 329. Department of Statistics University of Washington, 1998.

Fayyad U., and Irani K. B., The attribute selection problem in decision tree generation. In proceedings of Tenth National Conference on Artificial Intelligence, pp. 104–110, Cambridge, MA: AAAI Press/MIT Press, 1992.

Fayyad, U., Piatesky-Shapiro, G. & Smyth P., From Data Mining to Knowledge Discovery: An Overview. In U. Fayyad, G. Piatetsky-Shapiro, P. Smyth, & R. Uthurusamy (Eds), Advances in Knowledge Discovery and Data Mining, pp 1-30, AAAI/MIT Press, 1996.

Fayyad, U., Grinstein, G. and Wierse, A., Information Visualization in Data Mining and Knowledge Discovery, Morgan Kaufmann, 2001.

Feigenbaum E. (1988): *Knowledge Processing – From File Servers to Knowledge Servers. In J.R. Queinlan ed., "Applications of Expert Systems". Vol. 2, Turing Institute Press, Chpater 1, pp. 3-11*

Ferri C., Flach P., and Hernández-Orallo J., Learning Decision Trees Using the Area Under the ROC Curve. In Claude Sammut and Achim Hoffmann, editors, Proceedings of the 19th International Conference on Machine Learning, pp. 139-146. Morgan Kaufmann, July 2002

Fifield D. J., Distributed Tree Construction From Large Datasets, Bachelor's Honor Thesis, Australian National University, 1992.

Fisher,R.A., 1936, The use of multiple measurements in taxonomic problems" Annual Eugenics, 7, Part II, pp. 179-188.

Fisher, D., 1987, Knowledge acquisition via incremental conceptual clustering, in machine learning 2, pp. 139-172.

Fischer, B., "Decomposition of Time Series - Comparing Different Methods in Theory and Practice", Eurostat Working Paper, 1995.

Fix, E., and Hodges, J.L., Discriminatory analysis. Nonparametric discrimination. Consistency properties. Technical Report 4, US Air Force School of Aviation Medicine. Randolph Field, TX, 1957.

Fortier, J.J. and Solomon, H. 1996. Clustering procedures. In proceedings of the Multivariate Analysis, '66, P.R. Krishnaiah (Ed.), pp. 493-506.

Fountain, T. Dietterich T., Sudyka B., "Mining IC Test Data to Optimize VLSI Testing", ACM SIGKDD Conference, 2000, pp. 18-25, 2000.

Frank E., Hall M., Holmes G., Kirkby R., Pfahringer B., WEKA - A Machine Learning Workbench for Data Mining, in O. Maimon, L. Rokach, editors The Data Mining and Knowledge Discovery Handbook, Springer, pp. 1305-1314, 2005.

Frawley W. J., Piatetsky-Shapiro G., and Matheus C. J., "Knowledge Discovery in Databases: An Overview," G. Piatetsky-Shapiro and W. J. Frawley, editors, Knowledge Discovery in Databases, 1-27, AAAI Press, Menlo Park,

California, 1991.

Freitas A. (2005), "Evolutionary Algorithms for Data Mining", in Oded Maimon and Lior Rokach (Eds.), The Data Mining and Knowledge Discovery Handbook, Springer, pp. 435-467.

Freitas X., and Lavington S. H., Mining Very Large Databases With Parallel Processing, Kluwer Academic Publishers, 1998.

Frelicot C. and Mascarilla L., Reject Strategies Driver Combination of Pattern Classifiers, 2001.

Freund S. (1995), Boosting a weak learning algorithm by majority. Information and Computation, 121(2):256-285, 1995

Freund S. (2001), An adaptive version of the boost by majority algorithm, Machine Learning 43(3): 293-318.

Freund S., A more robust boosting algorithm, arXiv:0905.2138, 2009.

Yoav Freund and Llew Mason. The Alternating Decision Tree Algorithm. Proceedings of the 16th International Conference on Machine Learning, pages 124-133 (1999)

Freund, Y., Schapire, R.E., (1997), A decision-theoretic generalization of on-line learning and an application to boosting. Journal of Computer and System Sciences 1(55) 119–139.

Freund Y. and Schapire R. E., Experiments with a new boosting algorithm. In Machine Learning: Proceedings of the Thirteenth International Conference, pages 325-332, 1996.

Friedman, J.H. & Tukey, J.W., A Projection Pursuit Algorithm for Exploratory Data Analysis, IEEE Transactions on Computers, 23: 9, 881-889, 1973.

Friedman, J., Kohavi, R., Yun, Y. 1996. Lazy decision trees. Proceedings of the Thirteenth National Conference on Artificial Intelligence. (pp. 717-724). Cambridge, MA: AAAI Press/MIT Press.

Friedman N., Geiger D., and Goldszmidt M., Bayesian Network Classifiers, Machine Learning 29: 2-3, 131-163, 1997.

Friedman, J., T. Hastie and R. Tibshirani (2000) Additive Logistic Regression: a Statistical View of Boosting, Annals of Statistics, 28(2):337-407.

Friedman J. H., A recursive partitioning decision rule for nonparametric classifiers. IEEE Trans. on Comp., C26:404-408, 1977.

Friedman, J. H., "Multivariate Adaptive Regression Splines", The Annual Of Statistics, 19, 1-141, 1991.

Friedman, J.H. (1997a). Data Mining and Statistics: What is the connection? 1997.

Friedman, J.H. (1997b). On bias, variance, 0/1 - loss and the curse of dimensionality, Data Mining and Knowledge Discovery, 1: 1, 55-77, 1997.

Friedman, J.H., 2002. Stochastic gradient boosting. Comput. Statist. Data Anal. 38 (4), 367-378.

Fu Q., Hu S., and Zhao S. (2005), Clusterin-based selective neural network ensemble, Journal of Zhejiang University SCIENCE, 6A(5), 387-392.

Fukunaga, K., Introduction to Statistical Pattern Recognition. San Diego, CA: Academic, 1990.

Fürnkranz, J., More efficient windowing, In Proceeding of The 14th national

Conference on Artificial Intelegence (AAAI-97), pp. 509-514, Providence, RI. AAAI Press, 1997.

Gago, P. and Bentos, C. (1998). A metric for selection of the most promising rules. In *Proceedings of the 2nd European Conference on The Pronciples of Data Mining and Knowledge Discovery (PKDD'98)*.

Gallinari, P., Modular Neural Net Systems, Training of. In (Ed.) M.A. Arbib. The Handbook of Brain Theory and Neural Networks, Bradford Books/MIT Press, 1995.

Gama J., A Linear-Bayes Classifier. In C. Monard, editor, Advances on Artificial Intelligence – SBIA2000. LNAI 1952, pp 269-279, Springer Verlag, 2000

Gams, M., New Measurements Highlight the Importance of Redundant Knowledge. In European Working Session on Learning, Montpeiller, France, Pitman, 1989.

Garcia-Pddrajas N., Garcia-Osorio C., Fyfe C., Nonlinear Boosting Projections for Ensemble Construction, Journal of Machine Learning Research 8 (2007) 1-33.

Gardner M., Bieker, J., Data mining solves tough semiconductor manufacturing problems. KDD 2000: pp. 376-383, 2000.

Gehrke J., Ganti V., Ramakrishnan R., Loh W., BOAT-Optimistic Decision Tree Construction. SIGMOD Conference 1999: pp. 169-180, 1999.

Gehrke J., Ramakrishnan R., Ganti V., RainForest - A Framework for Fast Decision Tree Construction of Large Datasets,Data Mining and Knowledge Discovery, 4 (2/3) 127-162, 2000.

Gelfand S. B., Ravishankar C. S., and Delp E. J., An iterative growing and pruning algorithm for classification tree design. IEEE Transaction on Pattern Analysis and Machine Intelligence, 13(2):163-174, 1991.

Geman S., Bienenstock, E., and Doursat, R., Neural networks and the bias/variance dilemma. Neural Computation, 4:1-58, 1995.

George, E. and Foster, D. (2000),Calibration and empirical Bayes variable selection, Biometrika, 87(4):731-747.

Gey, S., Poggi, J.-M., 2006. Boosting and instability for regression trees Comput. Statist. Data Anal. 50, 533-550.

Ghani, R., (2000), Using error correcting output codes for text classification. In: Proceedings of the 17th International Conference on Machine Learning, Morgan Kaufmann 303–310.

Giacinto G., Roli F., and Fumera G., Design of effective multiple classifier systems by clustering of classifiers, in 15th International Conference on Pattern Recognition, ICPR 2000, pp. 160-163, September 2000.

Gilad-Bachrach, R., Navot, A. and Tisliby. (2004) N. Margin based feature selection - theory and algorithms. *Proceeding of the 21'st International Conferenc on Machine Learning*, 2004.

Gillo M. W., MAID: A Honeywell 600 program for an automatised survey analysis. Behavioral Science 17: 251-252, 1972.

Giraud–Carrier Ch., Vilalta R., Brazdil R., Introduction to the Special Issue of on Meta-Learning, Machine Learning, 54 (3), 197-194, 2004.

Gluck, M. and Corter, J. (1985). Information, uncertainty, and the utility of categories. Proceedings of the Seventh Annual Conference of the Cognitive Science Society (pp. 283-287). Irvine, California: Lawrence Erlbaum Associates.

Grossman R., Kasif S., Moore R., Rocke D., and Ullman J., Data mining research: Opportunities and challenges. Report of three NSF workshops on mining large, massive, and distributed data, 1999.

Grumbach S., Milo T., Towards Tractable Algebras for Bags. Journal of Computer and System Sciences 52(3): 570-588, 1996.

Guha, S., Rastogi, R. and Shim, K. CURE: An efficient clustering algorithm for large databases. In Proceedings of ACM SIGMOD International Conference on Management of Data, pages 73-84, New York, 1998.

Gunter S., Bunke H. , Feature Selection Algorithms for the generation of multiple classifier systems, Pattern Recognition Letters, 25(11):1323-1336, 2004.

Guo Y. and Sutiwaraphun J., Knowledge probing in distributed data mining, in Proc. 4h Int. Conf. Knowledge Discovery Data Mining, pp 61-69, 1998.

Guruswami, V., Sahai, A. (1999). Multiclass learning, boosting, and error-correcting codes. Proc. 12th Annual Conf. Computational Learning Theory (pp. 145155). Santa Cruz, California.

Guyon I. and Elisseeff A., "An introduction to variable and feature selection", Journal of Machine Learning Research 3, pp. 1157-1182, 2003.

Hall, M. Correlation- based Feature Selection for Machine Learning. University of Waikato, 1999.

Hampshire, J. B., and Waibel, A. The meta-Pi network - building distributed knowledge representations for robust multisource pattern-recognition. Pattern Analyses and Machine Intelligence 14(7): 751-769, 1992.

Han, J. and Kamber, M. Data Mining: Concepts and Techniques. Morgan Kaufmann Publishers, 2001.

Hancock T. R., Jiang T., Li M., Tromp J., Lower Bounds on Learning Decision Lists and Trees. Information and Computation 126(2): 114-122, 1996.

Hand, D., Data Mining – reaching beyond statistics, Research in Official Stat. 1(2):5-17, 1998.

Hansen, L. K., and Salamon, P., Neural network ensembles. IEEE Transactions on Pattern Analysis and Machine Intelligence, 12(10), 993-1001, 1990.

Hansen J., Combining Predictors. Meta Machine Learning Methods and Bias/Variance & Ambiguity Decompositions. PhD dissertation. Aurhus University. 2000.

Hartigan, J. A. Clustering algorithms. John Wiley and Sons., 1975.

Hastie, T., Tibshirani, R., (1998), Classification by pairwise coupling. The Annals of Statistics 2 451-471.

Huang, Z., Extensions to the k-means algorithm for clustering large data sets with categorical values. Data Mining and Knowledge Discovery, 2(3), 1998.

Haykin, S., (1999), Neural Networks - A Compreensive Foundation. 2nd edn. Prentice-Hall, New Jersey.

He D. W., Strege B., Tolle H., and Kusiak A., Decomposition in Automatic Generation of Petri Nets for Manufacturing System Control and Scheduling, International Journal of Production Research, 38(6): 1437-1457, 2000.

Hilderman, R. and Hamilton, H. (1999). Knowledge discovery and interesting-
    ness measures: A survey. In *Technical Report CS 99-04*. Department of
    Computer Science, University of Regina.

Ho T. K. , Hull J.J., Srihari S.N.,Decision Combination in Multiple Classifier
    Systems, PAMI 1994, 16(1):66–75.

Ho T. K., Nearest Neighbors in Random Subspaces, Proc. of the Second Interna-
    tional Workshop on Statistical Techniques in Pattern Recognition, Sydney,
    Australia, August 11-13, 1998, 640–648.

Ho T. K., The Random Subspace Method for Constructing Decision Forests,
    IEEE Transactions on Pattern Analysis and Machine Intelligence, Vol. 20,
    No. 8, 1998, pp. 832-844.

Ho T. K., Multiple Classifier Combination: Lessons and Next Steps, in Kandel
    and Bunke, (eds.), Hybrid Methods in Pattern Recognition, World Scien-
    tific, 2002, 171–198.

Holland, J.H., (1975), Adaptation in Natural and Artificial Systems. University
    of Michigan Press.

Holmes, G. and Nevill-Manning, C. G. (1995) . Feature selection via the discovery
    of simple classification rules. In *Proceedings of the Symposium on Intelligent
    Data Analysis*, Baden- Baden, Germany.

Holmstrom, L., Koistinen, P., Laaksonen, J., and Oja, E., Neural and statistical
    classifiers - taxonomy and a case study. IEEE Trans. on Neural Networks,
    8,:5–17, 1997.

Holte, R. C.; Acker, L. E.; and Porter, B. W., Concept learning and the problem
    of small disjuncts. In Proceedings of the 11th International Joint Conference
    on Artificial Intelligence, pp. 813-818, 1989.

Holte R. C., Very simple classification rules perform well on most commonly used
    datasets. Machine Learning, 11:63-90, 1993.

Hong S., Use of Contextual Information for Feature Ranking and Discretiza-
    tion, IEEE Transactions on Knowledge and Data Engineering, 9(5):718-730,
    1997.

Hoppner F. , Klawonn F., Kruse R., Runkler T., Fuzzy Cluster Analysis, Wiley,
    2000.

Hothorn T., Lausen B., Bundling classifiers by bagging trees, Computational
    Statistics and Data Analysis 49 (2005) 1068-1078

Hrycej T., Modular Learning in Neural Networks. New York: Wiley, 1992.

Hsu, C.W., Lin, C.J., (2002), A comparison of methods for multi-class support
    vector machines. IEEE Transactions on Neural Networks 13(2) 415–425.

Hu, X., Using Rough Sets Theory and Database Operations to Construct a Good
    Ensemble of Classifiers for Data Mining Applications. ICDM01. pp 233-240,
    2001.

Hu Q., Yu D., Xie Z., Li X., EROS: Ensemble rough subspaces,Pattern Recogni-
    tion 40 (2007) 3728 - 3739.

Hu Q. H., Yu D. R., Wang M. Y., Constructing Rough Decision Forests, D. Slezak
    et al. (Eds.): RSFDGrC 2005, LNAI 3642, Springer, 2005, pp. 147-156

Huang Y. S. and Suen C. Y. , A method of combining multiple experts for
    the recognition of unconstrained handwritten numerals, IEEE Trans. Patt.

Anal. Mach. Intell. 17 (1995) 90-94.

Hubert, L. and Arabie, P. (1985) Comparing partitions. Journal of Classification, 5. 193-218.

Hunter L., Klein T. E., Finding Relevant Biomolecular Features. ISMB 1993, pp. 190-197, 1993.

Hwang J., Lay S., and Lippman A., Nonparametric multivariate density estimation: A comparative study, IEEE Transaction on Signal Processing, 42(10): 2795-2810, 1994.

Hyafil L. and Rivest R.L., Constructing optimal binary decision trees is NP-complete. Information Processing Letters, 5(1):15-17, 1976.

Islam M. M., Yao X., Murase K., A constructive algorithm for training cooperative neuralnetwork ensembles, IEEE Transactions on Neural Networks 14 (4)(2003) 820-834.

Jackson, J., *A User's Guide to Principal Components*. New York: John Wiley and Sons, 1991.

Jacobs, R. A., Jordan, M. I., Nowlan, S. J., and Hinton, G. E. Adaptive mixtures of local experts. Neural Computation 3(1):79-87, 1991.

Jain, A. and Zonker, D., Feature Selection: Evaluation, Application, and Small Sample Performance. IEEE Trans. on Pattern Analysis and Machine Intelligence, 19, 153-158, 1997.

Jain, A.K. Murty, M.N. and Flynn, P.J. Data Clustering: A Survey. ACM Computing Surveys, Vol. 31, No. 3, September 1999.

Jang J., "Structure determination in fuzzy modeling: A fuzzy CART approach," in Proc. IEEE Conf. Fuzzy Systems, 1994, pp. 480485.

Janikow, C.Z., Fuzzy Decision Trees: Issues and Methods, IEEE Transactions on Systems, Man, and Cybernetics, Vol. 28, Issue 1, pp. 1-14. 1998.

Jenkins R. and Yuhas, B. P. A simplified neural network solution through problem decomposition: The case of Truck backer-upper, IEEE Transactions on Neural Networks 4(4):718-722, 1993.

Jimenez, L. O., & Landgrebe D. A., Supervised Classification in High- Dimensional Space: Geometrical, Statistical, and Asymptotical Properties of Multivariate Data. IEEE Transaction on Systems Man, and Cybernetics — Part C: Applications and Reviews, 28:39-54, 1998.

Johansen T. A. and Foss B. A., A narmax model representation for adaptive control based on local model -Modeling, Identification and Control, 13(1):25-39, 1992.

John G. H., and Langley P., Estimating Continuous Distributions in Bayesian Classifiers. Proceedings of the Eleventh Conference on Uncertainty in Artificial Intelligence. pp. 338-345. Morgan Kaufmann, San Mateo, 1995.

John G. H., Kohavi R., and Pfleger P., Irrelevant features and the subset selection problem. In Machine Learning: Proceedings of the Eleventh International Conference. Morgan Kaufmann, 1994.

John G. H., Robust linear discriminant trees. In D. Fisher and H. Lenz, editors, Learning From Data: Artificial Intelligence and Statistics V, Lecture Notes in Statistics, Chapter 36, pp. 375-385. Springer-Verlag, New York, 1996.

Jordan, M. I., and Jacobs, R. A. Hierarchies of adaptive experts. In Advances in

Neural Information Processing Systems, J. E. Moody, S. J. Hanson, and R. P. Lippmann, Eds., vol. 4, Morgan Kaufmann Publishers, Inc., pp. 985-992, 1992.

Jordan, M. I., and Jacobs, R. A., Hierarchical mixtures of experts and the EM algorithm. Neural Computation, 6, 181-214, 1994.

Joshi, V. M., "On Evaluating Performance of Classifiers for Rare Classes", Second IEEE International Conference on Data Mining, IEEE Computer Society Press, pp. 641-644, 2002.

Kamath, C., and E. Cantu-Paz, Creating ensembles of decision trees through sampling, Proceedings, 33-rd Symposium on the Interface of Computing Science and Statistics, Costa Mesa, CA, June 2001.

Kamath, C., Cant-Paz, E. and Littau, D. (2002). Approximate splitting for ensembles of trees using histograms. In Second SIAM International Conference on Data Mining (SDM-2002).

Kanal, L. N., "Patterns in Pattern Recognition: 1968-1974". IEEE Transactions on Information Theory IT-20, 6: 697-722, 1974.

Kang H., Lee S., Combination Of Multiple Classifiers By Minimizing The Upper Bound Of Bayes Error Rate For Unconstrained Handwritten Numeral Recognition, International Journal of Pattern Recognition and Artificial Intelligence, 19(3):395 - 413, 2005.

Kargupta, H. and Chan P., eds, Advances in Distributed and Parallel Knowledge Discovery , pp. 185-210, AAAI/MIT Press, 2000.

Kass G. V., An exploratory technique for investigating large quantities of categorical data. Applied Statistics, 29(2):119-127, 1980.

Kaufman, L. and Rousseeuw, P.J., 1987, Clustering by Means of Medoids, In Y. Dodge, editor, Statistical Data Analysis, based on the L1 Norm, pp. 405-416, Elsevier/North Holland, Amsterdam.

Kaufmann, L. and Rousseeuw, P.J. Finding groups in data. New-York: Wiley, 1990.

Kearns M. and Mansour Y., A fast, bottom-up decision tree pruning algorithm with near-optimal generalization, in J. Shavlik, ed., 'Machine Learning: Proceedings of the Fifteenth International Conference', Morgan Kaufmann Publishers, Inc., pp. 269-277, 1998.

Kearns M. and Mansour Y., On the boosting ability of top-down decision tree learning algorithms. Journal of Computer and Systems Sciences, 58(1): 109-128, 1999.

Kenney, J. F. and Keeping, E. S. "Moment-Generating and Characteristic Functions," "Some Examples of Moment-Generating Functions," and "Uniqueness Theorem for Characteristic Functions." §4.6-4.8 in Mathematics of Statistics, Pt. 2, 2nd ed. Princeton, NJ: Van Nostrand, pp. 72-77, 1951.

Kerber, R., 1992, ChiMerge: Discretization of numeric attributes, in AAAI-92, Proceedings Ninth National Conference on Artificial Intelligence, pp. 123-128, AAAI Press/MIT Press.

Kim J.O. & Mueller C.W., Factor Analysis: Statistical Methods and Practical Issues. Sage Publications, 1978.

Kim, D.J., Park, Y.W. and Park,. A novel validity index for determination of the optimal number of clusters. IEICE Trans. Inf., Vol. E84-D, no.2 (2001), 281-285.

King, B. Step-wise Clustering Procedures, J. Am. Stat. Assoc. 69, pp. 86-101, 1967.

Kira, K. and Rendell, L. A., A practical approach to feature selection. In *Machine Learning: Proceedings of the Ninth International Conference.*, 1992.

Klautau, A., Jevtić, N., Orlistky, A., (2003), On nearest-neighbor error-correcting output codes with application to all-pairs multiclass support vector machines. Journal of Machine Learning Research **4** 1-15.

Klosgen W. and Zytkow J. M., "KDD: The Purpose, Necessity and Chalanges", Klosgen W. and Zytkow J. M. (Eds.), Handbook of Data Mining and Knowledge Discovery, pp. 1-9. Oxford University Press, 2002.

Knerr, S., Personnaz, L., Dreyfus, G., (1992), Handwritten digit recognition by neural networks with single-layer training. IEEE Transactions on Neural Networks **3**(6) 962–968.

Knerr, S., Personnaz, L., Dreyfus, G., (1990), In: Single-layer learning revisited: a stepwise procedure for building and training a neural network. Springer-Verlag, pp. 41–50

Kohavi R. and John G., The Wrapper Approach, In Feature Extraction, Construction and Selection: A Data Mining Perspective, H. Liu and H. Motoda (eds.), Kluwer Academic Publishers, 1998.

Kohavi, R. and Kunz, C. (1997), Option decision trees with majority votes, in D. Fisher, ed., 'Machine Learning: Proceedings of the Fourteenth International Conference', Morgan Kaufmann Publishers, Inc., pp. 161–169.

Kohavi R., and Provost F., Glossary of Terms, Machine Learning 30(2/3): 271-274, 1998.

Kohavi R. and Quinlan J. R., Decision-tree discovery. In Klosgen W. and Zytkow J. M., editors, Handbook of Data Mining and Knowledge Discovery, chapter 16.1.3, pages 267-276. Oxford University Press, 2002.

Kohavi R. and Sommerfield D., Targeting business users with decision table classifiers, in R. Agrawal, P. Stolorz & G. Piatetsky-Shapiro, eds, 'Proceedings of the Fourth International Conference on Knowledge Discovery and Data Mining', AAAI Press, pp. 249-253, 1998.

Kohavi R. and Wolpert, D. H., Bias Plus Variance Decomposition for Zero-One Loss Functions, Machine Learning: Proceedings of the 13th International Conference. Morgan Kaufman, 1996.

Kohavi R., Becker B., and Sommerfield D., Improving simple Bayes. In Proceedings of the European Conference on Machine Learning, 1997.

Kohavi, R., Bottom-up induction of oblivious read-once decision graphs, in F. Bergadano and L. De Raedt, editors, Proc. European Conference on Machine Learning, pp. 154-169, Springer-Verlag, 1994.

Kohavi R., Scaling up the accuracy of naive-bayes classifiers: a decision-tree hybrid. In Proceedings of the Second International Conference on Knowledge Discovery and Data Mining, pages 114–119, 1996.

Kolcz, A. Chowdhury, and J. Alspector, Data duplication: An imbalance problem "In Workshop on Learning from Imbalanced Data Sets" (ICML), 2003.

Kolen, J. F., and Pollack, J. B., Back propagation is sesitive to initial conditions. In Advances in Neural Information Processing Systems, Vol. 3, pp. 860-867 San Francisco, CA. Morgan Kaufmann, 1991.

Koller, D. and Sahami, M. (1996). Towards optimal feature selection. In *Machine Learning: Proceedings of the Thirteenth International Conference on machine Learning*. Morgan Kaufmann, 1996.

Kolter, Z. J., Maloof, M. A. (2007). Dynamic Weighted Majority: An Ensemble Method. Journal of Machine Learning Research , 2756-2790.

Kong E. B. and Dietterich T. G., Error-correcting output coding corrects bias and variance. In Proc. 12th International Conference on Machine Learning, Morgan Kaufmann, CA, USA, 1995, pp. 313321.

Kononenko, I., Comparison of inductive and Naive Bayes learning approaches to automatic knowledge acquisition. In B. Wielinga (Ed.), Current Trends in Knowledge Acquisition, Amsterdam, The Netherlands IOS Press, 1990.

Kononenko, I., SemiNaive Bayes classifier, Proceedings of the Sixth European Working Session on Learning, pp. 206-219, Porto, Portugal: SpringerVerlag, 1991.

Kreβel, U., (1999), Pairwise classification and support vector machines. In Schölkopf, B., Burges, C.J.C., Smola, A.J., (Eds.), Advances in Kernel Methods - Support Vector Learning, MIT Press 185–208.

Krogh, A., and Vedelsby, J., Neural network ensembles, cross validation and active learning. In Advances in Neural Information Processing Systems 7, pp. 231-238 1995.

Krtowski M., Grze M., Global learning of decision trees by an evolutionary algorithm (Khalid Saeed and Jerzy Peja), Information Processing and Security Systems, Springer, pp. 401-410, 2005.

Krtowski M., An evolutionary algorithm for oblique decision tree induction, Proc. of ICAISC'04, Springer, LNCS 3070, pp.432-437, 2004.

Kuhn H. W., The Hungarian method for the assignment problem. Naval Research Logistics Quarterly, 2:83–97, 1955.

Kuncheva, L.I., (2005a), Using diversity measures for generating error-correcting output codes in classifier ensembles. Pattern Recognition Letters **26** 83–90.

Kuncheva L., Combining Pattern Classifiers, Wiley Press 2005.

Kuncheva, L., & Whitaker, C., Measures of diversity in classifier ensembles and their relationship with ensemble accuracy. Machine Learning, pp. 181–207, 2003.

Kuncheva L.I. (2005b) Diversity in multiple classifier systems (Editorial), Information Fusion, 6 (1), 2005, 3-4.

Kusiak A., Kurasek C., Data Mining of Printed-Circuit Board Defects, IEEE Transactions on Robotics and Automation, 17(2): 191-196, 2001.

Kusiak, E. Szczerbicki, and K. Park, A Novel Approach to Decomposition of Design Specifications and Search for Solutions, International Journal of Production Research, 29(7): 1391-1406, 1991.

Kusiak, A., Decomposition in Data Mining: An Industrial Case Study, IEEE Transactions on Electronics Packaging Manufacturing, Vol. 23, No. 4, pp. 345-353, 2000.

Kusiak, A., Rough Set Theory: A Data Mining Tool for Semiconductor Manufacturing, IEEE Transactions on Electronics Packaging Manufacturing, 24(1): 44-50, 2001A.

Kusiak, A., 2001, Feature Transformation Methods in Data Mining, IEEE Transactions on Elctronics Packaging Manufacturing, Vol. 24, No. 3, pp. 214–221, 2001B.

Lam L., Classifier combinations: implementations and theoretical issues. In J. Kittlerand F. Roli, editors, Multiple Classifier Systems, Vol. 1857 of Lecture Notes in ComputerScience, Cagliari, Italy, 2000, Springer, pp. 78-86.

Langdon W. B., Barrett S. J., Buxton B. F., Combining decision trees and neural networks for drug discovery, in: Genetic Programming, Proceedings of the 5th European Conference, EuroGP 2002, Kinsale, Ireland, 2002, pp. 60–70.

Langley, P. and Sage, S., Oblivious decision trees and abstract cases. in Working Notes of the AAAI-94 Workshop on Case-Based Reasoning, pp. 113-117, Seattle, WA: AAAI Press, 1994.

Langley, P. and Sage, S., Induction of selective Bayesian classifiers. in Proceedings of the Tenth Conference on Uncertainty in Artificial Intelligence, pp. 399-406. Seattle, WA: Morgan Kaufmann, 1994.

Langley, P., Selection of relevant features in machine learning, in Proceedings of the AAAI Fall Symposium on Relevance, pp. 140-144, AAAI Press, 1994.

Larsen, B. and Aone, C. 1999. Fast and effective text mining using linear-time document clustering. In Proceedings of the 5th ACM SIGKDD, 16-22, San Diego, CA.

Lazarevic A. and Obradovic Z., Effective pruning of neural network classifiers, in 2001 IEEE/INNS International Conference on Neural Networks, IJCNN 2001, pp. 796-801, July 2001.

Lee, S., Noisy Replication in Skewed Binary Classification, Computational Statistics and Data Analysis, 34, 2000.

Leigh W., Purvis R., Ragusa J. M., Forecasting the NYSE composite index with technical analysis, pattern recognizer, neural networks, and genetic algorithm: a case study in romantic decision support, Decision Support Systems 32(4): 361–377, 2002.

Lewis D., and Catlett J., Heterogeneous uncertainty sampling for supervised learning. In Machine Learning: Proceedings of the Eleventh Annual Conference, pp. 148-156 , New Brunswick, New Jersey, Morgan Kaufmann, 1994.

Lewis, D., and Gale, W., Training text classifiers by uncertainty sampling, In seventeenth annual international ACM SIGIR conference on research and development in information retrieval, pp. 3-12, 1994.

Jing Li, Nigel Allinson, Dacheng Tao, and Xuelong Li, Multitraining Support Vector Machine for Image Retrieval, IEEE Transactions on Image Processing, vol. 15, no. 11, pp. 3597-3601, November 2006.

Li X. and Dubes R. C., Tree classifier design with a Permutation statistic, Pattern Recognition 19:229-235, 1986.

Liao Y., and Moody J., Constructing Heterogeneous Committees via Input Feature Grouping, in Advances in Neural Information Processing Systems,

Vol.12, S.A. Solla, T.K. Leen and K.-R. Muller (Eds.),MIT Press, 2000.

Lim X., Loh W.Y., and Shih X., A comparison of prediction accuracy, complexity, and training time of thirty-three old and new classification algorithms . Machine Learning 40:203-228, 2000.

Lin Y. K. and Fu K., Automatic classification of cervical cells using a binary tree classifier. Pattern Recognition, 16(1):69-80, 1983.

Lin L., Wang X., Yeung D., Combining Multiple Classifiers Based On A Statistical Method For Handwritten Chinese Character Recognition, International Journal of Pattern Recognition and Artificial Intelligence, 19(8):1027 - 1040, 2005.

Lin H., Kao Y., Yang F., Wang P., Content-Based Image Retrieval Trained By Adaboost For Mobile Application, International Journal of Pattern Recognition and Artificial Intelligence, 20(4):525-541, 2006.

Lindbergh D.A.B. and Humphreys B.L., The Unified Medical Language System. In: van Bemmel JH and McCray AT, 1993 Yearbook of Medical Informatics. IMIA, the Nether-lands, page 41-51, 1993.

Ling C. X., Sheng V. S., Yang Q., Test Strategies for Cost-Sensitive Decision Trees IEEE Transactions on Knowledge and Data Engineering,18(8):1055-1067, 2006.

Liu C., Classifier combination based on confidence transformation, Pattern Recognition 38 (2005) 11 - 28

Liu H. & Motoda H., Feature Selection for Knowledge Discovery and Data Mining, Kluwer Academic Publishers, 1998.

Liu, H. and Setiono, R. (1996) A probabilistic approach to feature selection: A filter solution. In Machine Learning: *Proceedings of the Thirteenth International Conference on Machine Learning.* Morgan Kaufmann.

Liu, H., Hsu, W., and Chen, S. (1997). Using general impressions to analyze discovered classification rules. In *Proceedings of the Third International Conference on Knowledge Discovery and Data Mining (KDD'97).* Newport Beach, California.

Liu H., Mandvikar A., Mody J., An Empirical Study of Building Compact Ensembles. WAIM 2004: pp. 622-627.

Liu Y.: Generate Different Neural Networks by Negative Correlation Learning. ICNC (1) 2005: 149-156

Loh W.Y.,and Shih X., Split selection methods for classification trees. Statistica Sinica, 7: 815-840, 1997.

Loh W.Y. and Shih X., Families of splitting criteria for classification trees. Statistics and Computing 9:309-315, 1999.

Loh W.Y. and Vanichsetakul N., Tree-structured classification via generalized discriminant Analysis. Journal of the American Statistical Association, 83:715-728, 1988.

Long C., Bi-Decomposition of Function Sets Using Multi-Valued Logic, Eng.Doc. Dissertation, Technischen Universitat Bergakademie Freiberg 2003.

Lopez de Mantras R., A distance-based attribute selection measure for decision tree induction, Machine Learning 6:81-92, 1991.

Lorena, A.C., (2006),Investigação de estratégias para a geração de máquinas de vetores de suporte multiclasses [in portuguese], Ph.D. thesis, Departamento de Ciências de Computação, Instituto de Ciências Matemáticas e de Computação, Universidade de São Paulo, São Carlos, Brazil.

Lorena, A.C., Carvalho, A.C.P.L.F., Evolutionary design of multiclass support vector machines. Journal of Intelligent and Fuzzy Systems, 18(5): 445-454 (2007)

Lorena A. and de Carvalho A. C. P. L. F. : Evolutionary Design of Code-matrices for Multiclass Problems, in Oded Maimon and Lior Rokach (Eds.), Soft Computing for Knowledge Discovery and Data Mining, Springer, pp. 153-184, 2008.

Lu B.L., Ito M., Task Decomposition and Module Combination Based on Class Relations: A Modular Neural Network for Pattern Classification, IEEE Trans. on Neural Networks, 10(5):1244-1256, 1999.

Lu H., Setiono R., and Liu H., Effective Data Mining Using Neural Networks. IEEE Transactions on Knowledge and Data Engineering, 8 (6): 957-961, 1996.

Luba, T., Decomposition of multiple-valued functions, in Intl. Symposium on Multiple-Valued Logic', Bloomigton, Indiana, pp. 256-261, 1995.

Lubinsky D., Algorithmic speedups in growing classification trees by using an additive split criterion. Proc. AI&Statistics93, pp. 435-444, 1993.

Maher P. E. and Clair D. C,, Uncertain reasoning in an ID3 machine learning framework, in Proc. 2nd IEEE Int. Conf. Fuzzy Systems, 1993, pp. 712.

Maimon O., and Rokach, L. Data Mining by Attribute Decomposition with semi-conductors manufacturing case study, in Data Mining for Design and Manufacturing: Methods and Applications, D. Braha (ed.), Kluwer Academic Publishers, pp. 311-336, 2001.

Maimon O. and Rokach L., "Improving supervised learning by feature decomposition", Proceedings of the Second International Symposium on Foundations of Information and Knowledge Systems, Lecture Notes in Computer Science, Springer, pp. 178-196, 2002.

Maimon O., Rokach L., Ensemble of Decision Trees for Mining Manufacturing Data Sets, Machine Engineering, vol. 4 No1-2, 2004.

Maimon, O. and Rokach, L., Decomposition Methodology for Knowledge Discovery and Data Mining: Theory and Applications, Series in Machine Perception and Artificial Intelligence - Vol. 61, World Scientific Publishing, ISBN:981-256-079-3, 2005.

Mallows, C. L., Some comments on Cp . Technometrics 15, 661- 676, 1973.

Mangiameli P., West D., Rampal R., Model selection for medical diagnosis decision support systems, Decision Support Systems, 36(3): 247–259, 2004.

Mansour, Y. and McAllester, D., Generalization Bounds for Decision Trees, in Proceedings of the 13th Annual Conference on Computer Learning Theory, pp. 69-80, San Francisco, Morgan Kaufmann, 2000.

Marcotorchino, J.F. and Michaud, P. Optimisation en Analyse Ordinale des Donns. Masson, Paris.

Margineantu, D. (2001). Methods for Cost-Sensitive Learning. Doctoral Dissertation, Oregon State University.

Margineantu D. and Dietterich T., Pruning adaptive boosting. In Proc. Fourteenth Intl. Conf. Machine Learning, pages 211–218, 1997.

Martí, R., Laguna, M., Campos, V., (2005), Scatter search vs. genetic algorithms: An experimental evaluation with permutation problems. In Rego, C., Alidaee, B., eds.: Metaheuristic Optimization Via Adaptive Memory and Evolution: Tabu Search and Scatter Search. Kluwer Academic Publishers 263–282.

Martin J. K., An exact probability metric for decision tree splitting and stopping. An Exact Probability Metric for Decision Tree Splitting and Stopping, Machine Learning, 28 (2-3):257-291, 1997.

Martinez-Munoz G., Suarez A., Switching class labels to generate classification ensembles, Pattern Recognition, 38 (2005): 1483–1494.

Masulli, F., Valentini, G., (2000), Effectiveness of error correcting output codes in multiclass learning problems. In: Proceedings of the 1st International Workshop on Multiple Classifier Systems. Volume 1857 of Lecture Notes in Computer Science., Springer-Verlag 107–116.

Mayoraz, E., Alpaydim, E., (1998), Support vector machines for multi-class classification. Research Report IDIAP-RR-98-06, Dalle Molle Institute for Perceptual Artificial Intelligence, Martigny, Switzerland.

Mayoraz, E., Moreira, M., (1996), On the decomposition of polychotomies into dichotomies. Research Report 96-08, IDIAP, Dalle Molle Institute for Perceptive Artificial Intelligence, Martigny, Valais, Switzerland.

Mease D., Wyner W., Evidence Contrary to the Statistical View of Boosting, Journal of Machine Learning Research 9 (2008) 131-156

Mehta M., Rissanen J., Agrawal R., MDL-Based Decision Tree Pruning. KDD 1995: pp. 216-221, 1995.

Mehta M., Agrawal R. and Rissanen J., SLIQ: A fast scalable classifier for data mining: In Proc. If the fifth Int'l Conference on Extending Database Technology (EDBT), Avignon, France, March 1996.

Meir R., Ratsch G., An introduction to boosting and leveraging, In Advanced Lectures on Machine Learning, LNCS (2003), pp. 119-184.

Melville P., Mooney R. J., Constructing Diverse Classifier Ensembles using Artificial Training Examples. IJCAI 2003: 505-512

Menahem, E., Rokach, L., Elovici, Y., Troika - An Improved Stacking Schema for Classification Tasks, Information Sciences (to appear).

Menahem, E., Shabtai, A., Rokach, L., Elovici, Y., Improving malware detection by applying multi-inducer ensemble. Computational Statistics and Data Analysis, 53(4):1483–1494, 2009.

Meretakis, D. and Wthrich, B., Extending Nave Bayes Classifiers Using Long Itemsets, in Proceedings of the Fifth International Conference on Knowledge Discovery and Data Mining, pp. 165-174, San Diego, USA, 1999.

Merkwirth C., Mauser H., Schulz-Gasch T., Roche O., Stahl M., Lengauer T., Ensemble methods for classification in cheminformatics, Journal of Chemical Information and Modeling, 44(6):1971–1978, 2004.

Merler S., Caprile B., Furlanello C., Parallelizing AdaBoost by weights dynamics, Computational Statistics and Data Analysis 51 (2007) 2487-2498

Merz, C. J. and Murphy. P.M., UCI Repository of machine learning databases. Irvine, CA: University of California, Department of Information and Computer Science, 1998.

Merz, C. J., Using Correspondence Analysis to Combine Classifier, Machine Learning, 36(1-2):33-58, 1999.

Michalewicz, Z., Fogel, D.B., (2004), How to solve it: modern heuristics. Springer.

Michalski R. S., and Tecuci G.. Machine Learning, A Multistrategy Approach, Vol. J. Morgan Kaufmann, 1994.

Michalski R. S., A theory and methodology of inductive learning. Artificial Intelligence, 20:111- 161, 1983.

Michalski R. S., Understanding the nature of learning: issues and research directions, in R. Michalski, J. Carbonnel and T. Mitchell,eds, Machine Learning: An Artificial Intelligence Approach, Kaufmann, Paolo Alto, CA, pp. 3–25, 1986.

Michie D., Spiegelhalter D.J., Taylor C .C., Machine Learning, Neural and Statistical Classification, Prentice Hall, 1994.

Michie, D., Problem decomposition and the learning of skills, in Proceedings of the European Conference on Machine Learning, pp. 17-31, Springer-Verlag, 1995.

Mierswa I., Wurst M., Klinkenberg R., Scholz M., and Euler T.: YALE: Rapid Prototyping forComplex Data Mining Tasks, in Proceedings of the 12th ACM SIGKDDInternational Conference on Knowledge Discovery and Data Mining(KDD-06), 2006.

Mingers J., An empirical comparison of pruning methods for decision tree induction. Machine Learning, 4(2):227-243, 1989.

Minsky M., Logical vs. Analogical or Symbolic vs. Connectionist or Neat vs. Scruffy, in Artificial Intelligence at MIT., Expanding Frontiers, Patrick H. Winston (Ed.), Vol 1, MIT Press, 1990. Reprinted in AI Magazine, 1991.

Mishra, S. K. and Raghavan, V. V., An empirical study of the performance of heuristic methods for clustering. In Pattern Recognition in Practice, E. S. Gelsema and L. N. Kanal, Eds. 425436, 1994.

Mitchell, M., (1999), An introduction to Genetic Algorithms. MIT Press.

Mitchell, T., The need for biases in learning generalizations. Technical Report CBM-TR-117, Rutgers University, Department of Computer Science, New Brunswick, NJ, 1980.

Mitchell, T., Machine Learning, McGraw-Hill, 1997.

Montgomery D.C. (1997) Design and analysis, 4th edn. Wiley, New York.

Moody, J. and Darken, C., Fast learning in networks of locally tuned units. Neural Computations, 1(2):281-294, 1989.

Francisco Moreno-Seco, Jose M. Inesta, Pedro J. Ponce de Leon, and Luisa Mic, Comparison of Classifier Fusion Methods for Classification in Pattern Recognition Tasks, D. Y. Yeung et al. (Eds.): SSPR-SPR 2006, LNCS 4109, pp. 705–713, 2006.

Morgan J. N. and Messenger R. C., THAID: a sequential search program for the analysis of nominal scale dependent variables. Technical report, Institute for Social Research, Univ. of Michigan, Ann Arbor, MI, 1973.

Moskovitch R, Elovici Y, Rokach L, Detection of unknown computer worms based on behavioral classification of the host, Computational Statistics and Data Analysis, 52(9):4544–4566, 2008.

Muller W., and Wysotzki F., Automatic construction of decision trees for classification. Annals of Operations Research, 52:231-247, 1994.

Murphy, O. J., and McCraw, R. L. 1991. Designing storage efficient decision trees. IEEE-TC 40(3):315320.

Murtagh, F. A survey of recent advances in hierarchical clustering algorithms which use cluster centers. Comput. J. 26 354-359, 1984.

Murthy S. K., Kasif S., and Salzberg S.. A system for induction of oblique decision trees. Journal of Artificial Intelligence Research, 2:1-33, August 1994.

Murthy, S. and Salzberg, S. (1995), Lookahead and pathology in decision tree induction, in C. S. Mellish, ed., 'Proceedings of the 14th International Joint Con- ference on Articial Intelligence', Morgan Kaufmann, pp. 1025-1031.

Murthy S. K., Automatic Construction of Decision Trees from Data: A Multi-Disciplinary Survey. Data Mining and Knowledge Discovery, 2(4):345-389, 1998.

Myers E.W., An $O(ND)$ Difference Algorithm and Its Variations, Algorithmica, 1(1): page 251-266, 1986.

Naumov G.E., NP-completeness of problems of construction of optimal decision trees. Soviet Physics: Doklady, 36(4):270-271, 1991.

Neal R., Probabilistic inference using Markov Chain Monte Carlo methods. Tech. Rep. CRG-TR-93-1, Department of Computer Science, University of Toronto, Toronto, CA, 1993.

Ng, R. and Han, J. 1994. Very large data bases. In Proceedings of the 20th International Conference on Very Large Data Bases (VLDB94, Santiago, Chile, Sept.), VLDB Endowment, Berkeley, CA, 144155.

Niblett T. and Bratko I., Learning Decision Rules in Noisy Domains, Proc. Expert Systems 86, Cambridge: Cambridge University Press, 1986.

Niblett T., Constructing decision trees in noisy domains. In Proceedings of the Second European Working Session on Learning, pages 67-78, 1987.

Nowlan S. J., and Hinton G. E. Evaluation of adaptive mixtures of competing experts. In Advances in Neural Information Processing Systems, R. P. Lippmann, J. E. Moody, and D. S. Touretzky, Eds., vol. 3, pp. 774-780, Morgan Kaufmann Publishers Inc., 1991.

Nunez, M. (1988): *Economic induction: A case study.* In D. Sleeman (Ed.), Proceeding of the Third European Working Session on Learning. London: Pitman Publishing

Nunez, M. (1991): *The use of Background Knowledge in Decision Tree Induction.* Machine Learning, 6(1), pp. 231-250.

Oates, T., Jensen D., 1998, Large Datasets Lead to Overly Complex Models: An Explanation and a Solution, KDD 1998, pp. 294-298.

Ohno-Machado, L., and Musen, M. A. Modular neural networks for medical prognosis: Quantifying the benefits of combining neural networks for survival prediction. Connection Science 9, 1 (1997), 71-86.

Olaru C., Wehenkel L., A complete fuzzy decision tree technique, Fuzzy Sets and

Systems, 138(2):221–254, 2003.

Oliveira L.S., Sabourin R., Bortolozzi F., and Suen C. Y. (2003) A Methodology for Feature Selection using Multi-Objective Genetic Algorithms for Handwritten Digit String Recognition, *International Journal of Pattern Recognition and Artificial Intelligence*, 17(6):903-930.

Opitz, D., Feature Selection for Ensembles, In: Proc. 16th National Conf. on Artificial Intelligence, AAAI,1999, pp. 379-384.

Opitz, D. and Maclin, R., Popular Ensemble Methods: An Empirical Study, Journal of Artificial Research, 11: 169-198, 1999.

Opitz D. and Shavlik J., Generating accurate and diverse members of a neural-network ensemble. In David S. Touretzky, Michael C. Mozer, and Michael E. Hasselmo, editors, Advances in Neural Information Processing Systems, volume 8, pages 535–541. The MIT Press, 1996.

Pérez-Cruz, F., Artés-Rodríguez, A., (2002), Puncturing multi-class support vector machines. In: Proceedings of the 12th International Conference on Neural Networks (ICANN). Volume 2415 of Lecture Notes in Computer Science., Springer-Verlag 751–756.

Pagallo, G. and Huassler, D., Boolean feature discovery in empirical learning, Machine Learning, 5(1): 71-99, 1990.

S. Pang, D. Kim, S. Y. Bang, Membership authentication in the dynamic group by face classification using SVM ensemble. Pattern Recognition Letters, 24: 215–225, 2003.

Park C., Cho S., Evolutionary Computation for Optimal Ensemble Classifier in Lymphoma Cancer Classification. 521-530. Ning Zhong, Zbigniew W. Ras, Shusaku Tsumoto, Einoshin Suzuki (Eds.): Foundations of Intelligent Systems, 14th International Symposium, ISMIS 2003, Maebashi City, Japan, October 28-31, 2003, Proceedings. Lecture Notes in Computer Science, pp. 521-530, 2003.

Parmanto, B., Munro, P. W., and Doyle, H. R., Improving committee diagnosis with resampling techinques. In Touretzky, D. S., Mozer, M. C., and Hesselmo, M. E. (Eds). Advances in Neural Information Processing Systems, Vol. 8, pp. 882-888 Cambridge, MA. MIT Press, 1996.

Partridge D. , Yates W. B. (1996), Engineering multiversion neural-net systems, Neural Computation, 8(4):869-893.

Passerini, A., Pontil, M., Frasconi, P., (2004), New results on error correcting output codes of kernel machines. IEEE Transactions on Neural Networks 15 45–54.

Pazzani M., Merz C., Murphy P., Ali K., Hume T., and Brunk C. (1994): *Reducing Misclassification costs*. In Proc. 11th International conference on Machine Learning, 217-25. Morgan Kaufmann.

Pearl, J., Probabilistic Reasoning in Intelligent Systems: Networks of Plausible Inference. Morgan-Kaufmann, 1988.

Peng, F. and Jacobs R. A., and Tanner M. A., Bayesian Inference in Mixtures-of-Experts and Hierarchical Mixtures-of-Experts Models With an Application to Speech Recognition, Journal of the American Statistical Association 91, 953-960, 1996.

Peng Y., Intelligent condition monitoring using fuzzy inductive learning, Journal of Intelligent Manufacturing, 15 (3): 373-380, June 2004.

Perkowski, M.A., Luba, T., Grygiel, S., Kolsteren, M., Lisanke, R., Iliev, N., Burkey, P., Burns, M., Malvi, R., Stanley, C., Wang, Z., Wu, H., Yang, F., Zhou, S. and Zhang, J. S., Unified approach to functional decompositions of switching functions, Technical report, Warsaw University of Technology and Eindhoven University of Technology, 1995.

Perkowski, M., Jozwiak, L. and Mohamed, S., New approach to learning noisy Boolean functions, Proceedings of the Second International Conference on Computational Intelligence and Multimedia Applications, pp. 693–706, World Scientific, Australia, 1998.

Perkowski, M. A., A survey of literature on function decomposition, Technical report, GSRP Wright Laboratories, Ohio OH, 1995.

Perner P., Improving the Accuracy of Decision Tree Induction by Feature Pre-Selection, Applied Artificial Intelligence 2001, vol. 15, No. 8, p. 747-760.

Peterson W. W,, Weldon E. J., Error-correcting codes, The MIT Press; 2 edition, March 15, 1972.

Pfahringer, B., Bensusan H., and Giraud-Carrier C., Tell Me Who Can Learn You and I Can Tell You Who You are: Landmarking Various Learning Algorithms, In Proc. of the Seventeenth International Conference on Machine Learning (ICML2000), pages 743-750, 2000.

Pfahringer, B., Controlling constructive induction in CiPF, In Bergadano, F. and De Raedt, L. (Eds.), Proceedings of the seventh European Conference on Machine Learning, pp. 242-256, Springer-Verlag, 1994.

Pfahringer, B., Compression- based feature subset selection. In *Proceeding of the IJCAI- 95 Workshop on Data Engineering for Inductive Learning*, pp. 109-119, 1995.

Phama T., Smeuldersb A., Quadratic boosting, Pattern Recognition 41(2008): 331 - 341.

Piatetsky-Shapiro, G. (1991). *Discovery analysis and presentation of strong rules.* Knowledge Discovery in Databases, AAAI/MIT Press.

Pimenta, E., Gama, J., (2005), A study on error correcting output codes. In: Proceedings of the 2005 Portuguese Conference on Artificial Intelligence, IEEE Computer Society Press 218–223.

Poggio T., Girosi, F., Networks for Approximation and Learning, Proc. IEER, Vol 78(9): 1481-1496, Sept. 1990.

Polikar R., "Ensemble Based Systems in Decision Making," IEEECircuits and Systems Magazine, vol.6, no. 3, pp. 21-45, 2006.

Pratt, L. Y., Mostow, J., and Kamm C. A., Direct Transfer of Learned Information Among Neural Networks, in: Proceedings of the Ninth National Conference on Artificial Intelligence, Anaheim, CA, 584-589, 1991.

Prodromidis, A. L., Stolfo, S. J. and Chan, P. K., Effective and efficient pruning of metaclassifiers in a distributed data mining system. Technical report CUCS-017-99, Columbia Univ., 1999.

Provan, G. M. and Singh, M. (1996). Learning Bayesian networks using feature selection. In D. Fisher and H. Lenz, (Eds.), *Learning from Data, Lecture*

Notes in Statistics, pages 291– 300. Springer- Verlag, New York.

Provost, F. (1994): *Goal-Directed Inductive learning: Trading off accuracy for reduced error cost*. AAAI Spring Symposium on Goal-Driven Learning.

Provost F. and Fawcett T. (1997): *Analysis and visualization of Classifier Performance Comparison under Imprecise Class and Cost Distribution*. In Proceedings of KDD-97, pages 43-48. AAAI Press.

Provost F. and Fawcett T. (1998): *The case against accuracy estimation for comparing induction algorithms*. Proc. 15th Intl. Conf. On Machine Learning, pp. 445-453, Madison, WI.

Provost, F. and Fawcett, T. (2001), Robust {C}lassification for {I}mprecise {E}nvironments, Machine Learning, 42/3:203-231.

Provost, F.J. and Kolluri, V., A Survey of Methods for Scaling Up Inductive Learning Algorithms, Proc. 3rd International Conference on Knowledge Discovery and Data Mining, 1997.

Provost, F., Jensen, D. and Oates, T., 1999, Efficient Progressive Sampling, In Proceedings of the Fifth International Conference on Knowledge Discovery and Data Mining, pp.23-32.

Pujol, O., Tadeva, P., Vitrià, J., (2006), Discriminant ECOC: a heuristic method for application dependetn design of error correcting output codes. IEEE Transactions on Pattern Analysis and Machine Intelligence **28**(6) 1007–1012.

Quinlan, J. R. and Rivest, R. L., Inferring Decision Trees Using The Minimum Description Length Principle. Information and Computation, 80:227-248, 1989.

Quinlan, J.R. *Learning efficient classification procedures and their application to chess endgames*. R. Michalski, J. Carbonell, T. Mitchel. Machine learning: an AI approach. Los Altos, CA. Morgan Kaufman , 1983.

Quinlan, J.R., Induction of decision trees, Machine Learning 1, 81-106, 1986.

Quinlan, J.R., Simplifying decision trees, International Journal of Man-Machine Studies, 27, 221-234, 1987.

Quinlan, J.R., Decision Trees and Multivalued Attributes, J. Richards, ed., Machine Intelligence, V. 11, Oxford, England, Oxford Univ. Press, pp. 305-318, 1988.

Quinlan, J. R., Unknown attribute values in induction. In Segre, A. (Ed.), Proceedings of the Sixth International Machine Learning Workshop Cornell, New York. Morgan Kaufmann, 1989.

Quinlan, J. R., Unknown attribute values in induction. In Segre, A. (Ed.), Proceedings of the Sixth International Machine Learning Workshop Cornell, New York. Morgan Kaufmann, 1989.

Quinlan, J. R., C4.5: Programs for Machine Learning, Morgan Kaufmann, Los Altos, 1993.

Quinlan, J. R., Bagging, Boosting, and C4.5. In Proceedings of the Thirteenth National Conference on Artificial Intelligence, pages 725-730, 1996.

R Development Core Team (2004), R: A language and environment for statistical computing. R Foundation for Statistical Computing, Vienna, Austria. ISBN 3-900051-00-3, http://cran.r-project.org/, 2004

R'enyi A., Probability Theory, North-Holland, Amsterdam, 1970

Rätsch, G., Smola, A.J., Mika, S., (2003), Adapting codes and embeddings for polychotomies. In: Advances in Neural Information Processing Systems. Volume 15., MIT Press 513–520.

Ragavan, H. and Rendell, L., Look ahead feature construction for learning hard concepts. In Proceedings of the Tenth International Machine Learning Conference: pp. 252-259, Morgan Kaufman, 1993.

Rahman, A. F. R., and Fairhurst, M. C. A new hybrid approach in combining multiple experts to recognize handwritten numerals. Pattern Recognition Letters, 18: 781-790,1997.

Rakotomalala R., "TANAGRA: a free software for research andacademic purposes", in Proceedings of EGC'2005, RNTI-E-3, vol. 2,pp.697-702, 2005

Ramamurti, V., and Ghosh, J., Structurally Adaptive Modular Networks for Non-Stationary Environments, IEEE Transactions on Neural Networks, 10 (1):152-160, 1999.

Rand, W. M., Objective criteria for the evaluation of clustering methods. Journal of the American Statistical Association, 66: 846–850, 1971.

Rao, R., Gordon, D., and Spears, W., For every generalization action, is there really an equal or opposite reaction? Analysis of conservation law. In Proc. of the Twelveth International Conference on Machine Learning, pp. 471-479. Morgan Kaufmann, 1995.

Rastogi, R., and Shim, K., PUBLIC: A Decision Tree Classifier that Integrates Building and Pruning,Data Mining and Knowledge Discovery, 4(4):315-344, 2000.

Ratsch G., Onoda T., and Muller K. R., Soft Margins for Adaboost, Machine Learning 42(3):287-320, 2001.

Ray, S., and Turi, R.H. Determination of Number of Clusters in K-Means Clustering and Application in Color Image Segmentation. Monash university, 1999.

Buczak A. L. and Ziarko W., "Stages of The Discovery Process", Klosgen W. and Zytkow J. M. (Eds.), Handbook of Data Mining and Knowledge Discovery, pages 185-192. Oxford University Press, 2002.

Ridgeway, G., Madigan, D., Richardson, T. and O'Kane, J. (1998), "Interpretable Boosted Naive Bayes Classification", Proceedings of the Fourth International Conference on Knowledge Discovery and Data Mining, pp 101-104.

Rifkin, R., Klautau, A., (2004), In defense of one-vs-all classification. Journal of Machine Learning Research 5 1533–7928.

Rigoutsos I. and Floratos A., Combinatorial pattern discovery in biological sequences: The TEIRESIAS algorithm., Bioinformatics, 14(2): page 229, 1998.

Rissanen, J., Stochastic complexity and statistical inquiry. World Scientific, 1989.

Rodriguez J. J. (2006). Rotation Forest: A New Classifier Ensemble Method. IEEE Transactions on Pattern Analysis and Machine Intelligence, 20(10): 1619-1630

Rokach L., Ensemble Methods for Classifiers, in Oded Maimon and Lior Rokach (Eds.), The Data Mining and Knowledge Discovery Handbook, Springer, pp. 957-980, 2005.

Rokach L., Decomposition Methodology for Classification Tasks - A Meta Decomposer Framework, Pattern Analysis and Applications, 9(2006):257-271.

Rokach L., Genetic algorithm-based feature set partitioning for classification problems,Pattern Recognition, 41(5):1676–1700, 2008.

Rokach L., Mining manufacturing data using genetic algorithm-based feature set decomposition, Int. J. Intelligent Systems Technologies and Applications, 4(1):57-78, 2008.

Rokach, L., Collective-agreement-based pruning of ensembles. Computational Statistics and Data Analysis, 53(4):1015–1026, 2009.

Rokach L., Taxonomy for characterizing ensemble methods in classification tasks: A review and annotated bibliography, Computational Statistics and Data Analysis, 53(12):4046-4072, 2009.

Rokach L., Maimon O. and Lavi I., Space Decomposition In Data Mining: A Clustering Approach, Proceedings of the 14th International Symposium On Methodologies For Intelligent Systems, Maebashi, Japan, Lecture Notes in Computer Science, Springer-Verlag, 2003, pp. 24–31.

Rokach L., Averbuch M. and Maimon O., Information Retrieval System for Medical Narrative Reports, Lecture Notes in Artificial intelligence 3055, page 217-228 Springer-Verlag, 2004.

Rokach L., Maimon O., Arad O., "Improving Supervised Learning by Sample Decomposition", International Journal of Computational Intelligence and Applications, 5(1):37-54, 2005.

Rokach L., R. Arbel, O. Maimon, "Selective Voting - Getting More For Less in Sensor Fusion", International Journal of PatternRecognition and Artificial Intelligence, 20(3):329-350, 2006.

Rokach L., Chizi B., Maimon O., A Methodology for Improving the Performance of Non-ranker Feature Selection Filters, International Journal of Pattern Recognition and Artificial Intelligence, 21(5): 809-830, 2007.

Rokach L., Romano R., Maimon O., Negation Recognition in Medical Narrative Reports, Information Retrieval, 11(6): 499-538, 2008

Rokach L. and Maimon O., "Theory and Application of Attribute Decomposition", Proceedings of the First IEEE International Conference on Data Mining, IEEE Computer Society Press, pp. 473-480, 2001

Rokach L. and Maimon O., Top Down Induction of Decision Trees Classifiers: A Survey, IEEE SMC Transactions Part C. Volume 35, Number 3, 2005a.

Rokach L. and Maimon O., Feature Set Decomposition for Decision Trees, Journal of Intelligent Data Analysis, Volume 9, Number 2, 2005b, pp 131-158.

Rokach, L. and Maimon, O., Clustering methods, Data Mining and Knowledge Discovery Handbook, pp. 321–352, 2005, Springer.

Rokach, L. and Maimon, O., Data mining for improving the quality of manufacturing: a feature set decomposition approach, Journal of Intelligent Manufacturing, 17(3):285–299, 2006, Springer.

Rokach, L., Maimon, O., Data Mining with Decision Trees: Theory and Applications, World Scientific Publishing, 2008.

Ronco, E., Gollee, H., and Gawthrop, P. J., Modular neural network and self-decomposition. CSC Research Report CSC-96012, Centre for Systems and Control, University of Glasgow, 1996.

Rosen B. E., Ensemble Learning Using Decorrelated Neural Networks. Connect. Sci. 8(3): 373-384 (1996)

Rounds, E., A combined non-parametric approach to feature selection and binary decision tree design, Pattern Recognition 12, 313-317, 1980.

Rudin C., Daubechies I., and Schapire R. E., The Dynamics of Adaboost: Cyclic behavior and convergence of margins, Journal of Machine Learning Research Vol. 5, 1557-1595, 2004.

Rumelhart, D., G. Hinton and R. Williams, Learning internal representations through error propagation. In Parallel Distributed Processing: Explorations in the Microstructure of Cognition, Volume 1: Foundations, D. Rumelhart and J. McClelland (eds.) Cambridge, MA: MIT Press., pp 2540, 1986.

Saaty, X., The analytic hierarchy process: A 1993 overview. Central European Journal for Operations Research and Economics, Vol. 2, No. 2, p. 119-137, 1993.

Safavin S. R. and Landgrebe, D., A survey of decision tree classifier methodology. IEEE Trans. on Systems, Man and Cybernetics, 21(3):660-674, 1991.

Sakar A., Mammone R.J., Growing and pruning neural tree networks, IEEE Trans. on Computers 42, 291-299, 1993.

Salzberg. S. L., On Comparing Classifiers: Pitfalls to Avoid and a Recommended Approach. Data Mining and Knowledge Discovery, 1: 312-327, Kluwer Academic Publishers, Bosto, 1997.

Samuel, A., Some studies in machine learning using the game of checkers II: Recent progress. IBM J. Res. Develop., 11:601-617, 1967.

Schaffer, C., When does overfitting decrease prediction accuracy in induced decision trees and rule sets? In Proceedings of the European Working Session on Learning (EWSL-91), pp. 192-205, Berlin, 1991.

Schaffer, C., Selecting a classification method by cross-validation. Machine Learning 13(1):135-143, 1993.

Schaffer J., A Conservation Law for Generalization Performance. In Proceedings of the 11th International Conference on Machine Learning: pp. 259-265, 1993.

Schapire, R.E., *The Strength of Weak Learnability.* Machine learning 5(2), 197-227, 1990.

Schapire, R. E. (1997). Using output codes to boost multiclass learning problems. Proc. 14th Intl Conf. Machine Learning (pp. 313321). Nashville, TN, USA.

Schclar A., Rokach L.: Random Projection Ensemble Classifiers. ICEIS 2009: 309-316.

Schclar A., Rokach L., A. Meisels, Ensemble Methods for Improving the Performance of Neighborhood-based Collaborative Filtering, Proc. ACM RecSys 2009 (to appear).

Schlimmer, J. C. , Efficiently inducing determinations: A complete and systematic search algorithm that uses optimal pruning. In Proceedings of the 1993 International Conference on Machine Learning: pp 284-290, San Mateo, CA, Morgan Kaufmann, 1993.

Schmitt , M., On the complexity of computing and learning with multiplicative neural networks, Neural Computation 14: 2, 241-301, 2002.

Seewald, A. K., Exploring the Parameter State Space of Stacking. In: Proc. of the 2002 IEEE Int. Conf. on Data Mining, pp. 685–688, 2002A.

Seewald A.K., How to Make Stacking Better and Faster While Also Taking Care of an Unknown Weakness. In: Nineteenth International Conference on Machine Learning, 554–561, 2002B.

Seewald A.K., Towards Understanding Stacking. PhD Thesis, Vienna University of Technology, 2003.

Seewald, A.K. and Fürnkranz, J., Grading classifiers, Austrian research institute for Artificial intelligence, 2001.

Selfridge, O. G. Pandemonium: a paradigm for learning. In Mechanization of Thought Processes: Proceedings of a Symposium Held at the National Physical Laboratory, November, 1958, 513-526. London: H.M.S.O., 1958.

Selim, S. Z. AND Al-Sultan, K. 1991. A simulated annealing algorithm for the clustering problem. Pattern Recogn. 24, 10 (1991), 10031008.

Selim, S.Z., and Ismail, M.A. K-means-type algorithms: a generalized convergence theorem and characterization of local optimality. In IEEE transactions on pattern analysis and machine learning, vol. PAMI-6, no. 1, January, 1984.

Servedio, R., On Learning Monotone DNF under Product Distributions. Information and Computation 193, pp. 57-74, 2004.

Sethi, K., and Yoo, J. H., Design of multicategory, multifeature split decision trees using perceptron learning. Pattern Recognition, 27(7):939-947, 1994.

Sexton J., Laake P., LogitBoost with errors-in-variables, Computational Statistics and Data Analysis 52 (2008) 2549-2559

Shafer, J. C., Agrawal, R. and Mehta, M. , SPRINT: A Scalable Parallel Classifier for Data Mining, Proc. 22nd Int. Conf. Very Large Databases, T. M. Vijayaraman and Alejandro P. Buchmann and C. Mohan and Nandlal L. Sarda (eds), 544-555, Morgan Kaufmann, 1996.

Shapiro, A. D. and Niblett, T., Automatic induction of classification rules for a chess endgame, in M. R. B. Clarke, ed., Advances in Computer Chess 3, Pergamon, Oxford, pp. 73-92, 1982.

Shapiro, A. D., Structured induction in expert systems, Turing Institute Press in association with Addison-Wesley Publishing Company, 1987.

Sharkey A., Sharkey N., Combining diverse neural networks, The Knowledge Engineering Review 12(3): 231–247, 1997.

Sharkey, A., On combining artificial neural nets, Connection Science, Vol. 8, pp.299-313, 1996.

Sharkey, A., Multi-Net Iystems, In Sharkey A. (Ed.) Combining Artificial Neural Networks: Ensemble and Modular Multi-Net Systems. pp. 1-30, Springer-Verlag, 1999.

Shen, L., Tan, E.C., (2005), Seeking better output-codes with genetic algorithm for multiclass cancer classification. Submitted to Bioinformatics.

Shilen, S., Multiple binary tree classifiers. Pattern Recognition 23(7): 757-763, 1990.

Shilen, S., Nonparametric classification using matched binary decision trees. Pattern Recognition Letters 13: 83-87, 1992.

Simn, M.D.J., Pulido, J.A.G., Rodrguez, M.A.V., (2006), Prez, J.M.S., Criado, J.M.G., A genetic algorithm to design error correcting codes. In: Proceedings of the 13th IEEE Mediterranean Eletrotechnical Conference 2006, IEEE Computer Society Press 807–810.

Sivalingam D., Pandian N., Ben-Arie J., Minimal Classification Method With Error-Correcting Codes For Multiclass Recognition, International Journal of Pattern Recognition and Artificial Intelligence 19(5): 663 - 680, 2005.

Sklansky, J. and Wassel, G. N., Pattern classifiers and trainable machines. SpringerVerlag, New York, 1981.

Skurichina M. and Duin R.P.W., Bagging, boosting and the random subspace method for linear classifiers. Pattern Analysis and Applications, 5(2):121–135, 2002

Sloane N.J.A. (2007) A library of orthogonal arrays.

Smyth, P. and Goodman, R. (1991). *Rule induction using information theory.* Knowledge Discovery in Databases, AAAI/MIT Press.

Sneath, P., and Sokal, R. Numerical Taxonomy. W.H. Freeman Co., San Francisco, CA, 1973.

Snedecor, G. and Cochran, W. (1989). *Statistical Methods.* owa State University Press, Ames, IA, 8th Edition.

Sohn S. Y., Choi, H., Ensemble based on Data Envelopment Analysis, ECML Meta Learning workshop, Sep. 4, 2001.

Sohna S.Y., Shinb H.W., Experimental study for thecomparison of classifier combination methods, Pattern Recognition40 (2007) 33–40.

van Someren M.,Torres C. and Verdenius F. (1997): *A systematic Description of Greedy Optimisation Algorithms for Cost Sensitive Generalization.* X. Liu, P.Cohen, M. Berthold (Eds.): "Advance in Intelligent Data Analysis" (IDA-97) LNCS 1280, pp. 247-257.

Sonquist, J. A., Baker E. L., and Morgan, J. N., Searching for Structure. Institute for Social Research, Univ. of Michigan, Ann Arbor, MI, 1971.

Spirtes, P., Glymour C., and Scheines, R., Causation, Prediction, and Search. Springer Verlag, 1993.

Statnikov, A., Aliferis, C.F., Tsamardinos, I., (2005), Hardin, D., Levy, S., A comprehensive evaluation of multicategory methods for microarray gene expression cancer diagnosis. Bioinformatics 21(5) 631–643.

Steuer R.E.,Multiple Criteria Optimization: Theory, Computation and Application. John Wiley, New York, 1986.

Strehl A. and Ghosh J., Clustering Guidance and Quality Evaluation Using Relationship-based Visualization, Proceedings of Intelligent Engineering Systems Through Artificial Neural Networks, 5-8 November 2000, St. Louis, Missouri, USA, pp 483-488.

Strehl, A., Ghosh, J., Mooney, R.: Impact of similarity measures on web-page clustering. In Proc. AAAI Workshop on AI for Web Search, pp 58–64, 2000.

Sun Y., Todorovic S., Li L. Reducing The Overfitting Of Adaboost By Controlling Its Data Distribution Skewness, International Journal of Pattern

Recognition and Artificial Intelligence, 20(7):1093-1116, 2006.

Sun Y., Todorovic S., Li J., Wu D., Unifying the Error-Correcting and Output-Code AdaBoost within the Margin Framework, Proceedings of the 22nd international conference on Machine learning (2005), pp. 872-879.

Tadepalli, P. and Russell, S., Learning from examples and membership queries with structured determinations, Machine Learning, 32(3), pp. 245-295, 1998.

Tan A. C., Gilbert D., Deville Y., Multi-class Protein Fold Classification using a New Ensemble Machine Learning Approach. Genome Informatics, 14:206–217, 2003.

Tang E. K., Suganthan P. N., Yao X., An analysis of diversity measures, Machine Learning (2006) 65:247271

Tani T. and Sakoda M., Fuzzy modeling by ID3 algorithm and its application to prediction of heater outlet temperature, Proc. IEEE Internat. Conf. on Fuzzy Systems, March 1992, pp. 923-930.

Dacheng Tao and Xiaoou Tang, SVM-based Relevance Feedback Using Random Subspace Method, IEEE International Conference on Multimedia and Expo, pp. 647-652, 2004

Dacheng Tao, Xiaoou Tang, Xuelong Li, and Xindong Wu, Asymmetric Bagging and Random Subspace for Support Vector Machines-based Relevance Feedback in Image Retrieval, IEEE Transactions on Pattern Analysis and Machine Intelligence, vol. 28, no.7, pp. 1088-1099, July 2006

Dacheng Tao, Xuelong Li, and Stephen J. Maybank, Negative Samples Analysis in Relevance Feedback, IEEE Transactions on Knowledge and Data Engineering, vol. 19, no. 4, pp. 568-580, April 2007.

Tapia, E., González, J.C., García-Villalba, J., Villena, J., (2001), Recursive adaptive ECOC models. In: Proceedings of the 10th Portuguese Conference on Artificial Intelligence. Volume 2258 of Lecture Notes in Artificial Intelligence., Springer-Verlag 96–103.

Tapia, E., González, J.C., García-Villalba, J., (2003), Good error correcting output codes for adaptive multiclass learning. In: Proceedings of the 4th International Workshop on Multiple Classifier Systems 2003. Volume 2709 of Lecture Notes in Computer Science., Springer-Verlag 156–165.

Taylor P. C., and Silverman, B. W., Block diagrams and splitting criteria for classification trees. Statistics and Computing, 3(4):147-161, 1993.

Tibshirani, R., Walther, G. and Hastie, T. (2000). Estimating the number of clusters in a dataset via the gap statistic. Tech. Rep. 208, Dept. of Statistics, Stanford University.

Ting K.M. and Witten I.H. (1999), Issues in stacked generalization, J. Artif. Intell. Res. 10: 271289, 1999.

Towell, G. Shavlik, J., Knowledge-based artificial neural networks, Artificial Intelligence, 70: 119-165, 1994.

Tresp, V. and Taniguchi, M. Combining estimators using non-constant weighting functions. In Tesauro, G., Touretzky, D., & Leen, T. (Eds.), Advances in Neural Information Processing Systems, volume 7: pp. 419-426, The MIT Press, 1995.

Tsallis C., Possible Generalization of Boltzmann-Gibbs Statistics, J. Stat.Phys., 52, 479-487, 1988.

Tsao, C.A., Chang, Y.I., 2007. A stochastic approximation view of boosting. Comput. Stat. Data Anal. 52 (1), 325-344.

Tsoumakas G., Partalas I., Vlahavas I., A Taxonomy and Short Review of Ensemble Selection, Proc. Workshop on Supervised and Unsupervised Ensemble Methods, ECAI, Patras, Greece, 2008.

Tsymbal A., and Puuronen S., Ensemble Feature Selection with the Simple Bayesian Classification in Medical Diagnostics, In: Proc. 15thIEEE Symp. on Computer-Based Medical Systems CBMS2002, Maribor, Slovenia,IEEE CS Press, 2002, pp. 225-230.

Tsymbal A., and Puuronen S., and D. Patterson, Feature Selection for Ensembles of Simple Bayesian Classifiers,In: Foundations of Intelligent Systems: ISMIS2002, LNAI, Vol. 2366, Springer, 2002, pp. 592-600

Tsymbal A., Pechenizkiy M., Cunningham P., Diversity in search strategies for ensemble feature selection. Information Fusion 6(1): 83-98, 2005.

Tukey J.W., Exploratory data analysis, Addison-Wesley, Reading, Mass, 1977.

Tumer, K. and Ghosh J., Error Correlation and Error Reduction in Ensemble Classifiers, Connection Science, Special issue on combining artificial neural networks: ensemble approaches, 8 (3-4): 385-404, 1996.

Tumer, K., and Ghosh J., Linear and Order Statistics Combiners for Pattern Classification, in Combining Articial Neural Nets, A. Sharkey (Ed.), pp. 127-162, Springer-Verlag, 1999.

Tumer, K., and Ghosh J., Robust Order Statistics based Ensembles for Distributed Data Mining. In Kargupta, H. and Chan P., eds, Advances in Distributed and Parallel Knowledge Discovery , pp. 185-210, AAAI/MIT Press, 2000.

K. Tumer, C. N. Oza, Input decimated ensembles. Pattern Analysis and Application 6 (2003) 65-77.

Turney P. (1995): *Cost-Sensitive Classification: Empirical Evaluation of Hybrid Genetic Decision Tree Induction Algorithm.* Journal of Artificial Intelligence Research 2, pp. 369-409.

Turney P. (2000): *Types of Cost in Inductive Concept Learning.* Workshop on Cost Sensitive Learning at the 17th ICML, Stanford, CA.

Tutz, G., Binder, H., 2006. Boosting ridge regression. Computational Statistics and Data Analysis. Corrected Proof, Available online 22 December 2006, in press.

Tuv, E. and Torkkola, K., Feature filtering with ensembles using artificial contrasts. In *Proceedings of the SIAM 2005 Int. Workshop on Feature Selection for Data Mining*, Newport Beach, CA, 69-71, 2005.

Tyron R. C. and Bailey D.E. Cluster Analysis. McGraw-Hill, 1970.

Urquhart, R. Graph-theoretical clustering, based on limited neighborhood sets. Pattern recognition, vol. 15, pp. 173-187, 1982.

Utgoff, P. E., and Clouse, J. A., A Kolmogorov-Smirnoff Metric for Decision Tree Induction, Technical Report 96-3, University of Massachusetts, Department of Computer Science, Amherst, MA, 1996.

Utgoff, P. E., Perceptron trees: A case study in hybrid concept representations. Connection Science, 1(4):377-391, 1989.

Utgoff, P. E., Incremental induction of decision trees. Machine Learning, 4:161-186, 1989.

Utgoff, P. E., Decision tree induction based on efficient tree restructuring, Machine Learning 29 (1):5-44, 1997.

Vafaie, H. and De Jong, K. (1995). Genetic algorithms as a tool for restructuring feature space representations. In *Proceedings of the International Conference on Tools with A. I.* IEEE Computer Society Press.

Valentini G. and Masulli F., Ensembles of learning machines. In R. Tagliaferri andM. Marinaro, editors, Neural Nets, WIRN, Vol. 2486 of Lecture Notes in ComputerScience, Springer, 2002, pp. 3-19.

Valiant, L. G. (1984). A theory of the learnable. Communications of the ACM 1984, pp. 1134-1142.

Van Rijsbergen, C. J., Information Retrieval. Butterworth, ISBN 0-408-70929-4, 1979.

Van Zant, P., Microchip fabrication: a practical guide to semiconductor processing, New York: McGraw-Hill, 1997.

Vapnik, V.N., The Nature of Statistical Learning Theory. Springer-Verlag, New York, 1995.

Veyssieres, M.P. and Plant, R.E. Identification of vegetation state-and-transition domains in California's hardwood rangelands. University of California, 1998.

Vilalta R., Giraud–Carrier C., Brazdil P., "Meta-Learning", in O. Maimon and L. Rokach (Eds.), Handbook of Data Mining and Knowledge Discovery in Databases, pp. 731-748, Springer, 2005.

Villalba Santiago D., Rodrguez Juan J., Alonso Carlos J., An Empirical Comparison of Boosting Methods via OAIDTB, an Extensible Java Class Library, In II International Workshop on Practical Applications of Agents and Multiagent Systems - IWPAAMS'2003.

Wallace C. S. and Dowe D. L., Intrinsic classification by mml – the snob program. In Proceedings of the 7th Australian Joint Conference on Artificial Intelligence, pages 37-44, 1994.

Wallace, C. S., and Patrick J., Coding decision trees, Machine Learning 11: 7-22, 1993.

Wallace, C. S., MML Inference of Predictive Trees, Graphs and Nets. In A. Gammerman (ed), Computational Learning and Probabilistic Reasoning, pp 43-66, Wiley, 1996.

Wallet, B.C., Marchette, D.J., Solka, J.L., (1996), A matrix representation for genetic algorithms. In: Automatic object recognition VI, Proceedings of the International Society for Optical Engineering. 206–214.

Wallis, J.L., Houghten, S.K., (2002), A comparative study of search techniques applied to the minimum distance problem of BCH codes. Technical Report CS-02-08, Department of Computer Science, Brock University.

Walsh P., Cunningham P., Rothenberg S., O'Doherty S., Hoey H., Healy R., An artificial neural network ensemble to predict disposition and length of stay in children presenting with bronchiolitis. European Journal of Emergency Medicine. 11(5):259-264, 2004.

Wan, W. and Perkowski, M. A., A new approach to the decomposition of incompletely specified functions based on graph-coloring and local transformations and its application to FPGAmapping, In Proc. of the IEEE EURO-DAC '92, pp. 230-235, 1992.

Wanas Nayer M., Dara Rozita A. , Kamel Mohamed S., Adaptivefusion and cooperative training for classifier ensembles, PatternRecognition 39 (2006) 1781 - 1794

Wang, X. and Yu, Q. Estimate the number of clusters in web documents via gap statistic. May 2001.

Wang W., Jones P., Partridge D., Diversity between neural networks and decision trees for building multiple classifier systems, in: Proc. Int. Workshop on Multiple Classifier Systems (LNCS 1857), Springer, Calgiari, Italy, 2000, pp. 240–249.

Ward, J. H. Hierarchical grouping to optimize an objective function. Journal of the American Statistical Association, 58:236-244, 1963.

Warshall S., A theorem on Boolean matrices, Journal of the ACM 9, 1112, 1962.

Webb G., and Zheng Z., Multistrategy Ensemble Learning: Reducing Error by Combining Ensemble Learning Techniques. IEEE Transactions on Knowledge and Data Engineering, 16 No. 8:980-991, 2004.

Webb G., MultiBoosting: A technique for combining boosting and wagging. Machine Learning, 40(2): 159-196, 2000.

Weigend, A. S., Mangeas, M., and Srivastava, A. N. Nonlinear gated experts for time-series - discovering regimes and avoiding overfitting. International Journal of Neural Systems 6(5):373-399, 1995.

J. Weston and C. Watkins. Support vector machines for multi-class pattern recognition. In M. Verleysen (ed.) Proceedings of the 7th European Symposium on Artificial Neural Networks (ESANN-99), pp. 219224, Bruges, Belgium, 1999.

Widmer, G. and Kubat, M., 1996, Learning in the Presence of Concept Drift and Hidden Contexts, Machine Learning 23(1), pp. 69101.

Windeatt T. and Ardeshir G., An Empirical Comparison of Pruning Methods for Ensemble Classifiers, IDA2001, LNCS 2189, pp. 208-217, 2001.

Windeatt, T., Ghaderi, R., (2003), Coding and decoding strategies for multi-class learning problems. Information Fusion 4(1) 11–21.

Wolf L., Shashua A., Feature Selection for Unsupervised and Supervised Inference: The Emergence of Sparsity in a Weight-Based Approach, Journal of Machine Learning Research, Vol 6, pp. 1855-1887, 2005.

Wolpert, D., Macready, W. 1996. Combining Stacking with Bagging to Improve a Learning Algorithm. Santa Fe Institute Technical Report 96-03-123.

Wolpert, D.H., Stacked Generalization, Neural Networks, Vol. 5, pp. 241-259, Pergamon Press, 1992.

Wolpert, D. H., The relationship between PAC, the statistical physics framework, the Bayesian framework, and the VC framework. In D. H. Wolpert, editor, The Mathematics of Generalization, The SFI Studies in the Sciences of

Complexity, pages 117-214. AddisonWesley, 1995.

Wolpert, D. H., "The lack of a priori distinctions between learning algorithms," Neural Computation 8: 1341–1390, 1996.

Wolpert, D. H., "The supervised learning no-free-lunch theorems", Proceedings of the 6th Online World Conference on Soft Computing in Industrial Applications, 2001.

Woods K., Kegelmeyer W., Bowyer K., Combination of multiple classifiers using local accuracy estimates, IEEE Transactions on Pattern Analysis and Machine Intelligence 19:405–410, 1997.

Q. X. Wu , D. Bell and M. McGinnity, Multi-knowledge for decision making, Knowledge and Information Systems, 7(2005): 246-266

Wyse, N., Dubes, R. and Jain, A.K., A critical evaluation of intrinsic dimensionality algorithms, Pattern Recognition in Practice, E.S. Gelsema and L.N. Kanal (eds.), North-Holland, pp. 415–425, 1980.

Xu L., Krzyzak A., Suen C.Y., Methods of combining multiple classifiers and their application to handwriting recognition, IEEE Trans. SMC 22, 418-435, 1992

Yanim S., Kamel Mohamad S., Wong Andrew K.C., Wang Yang (2007). Cost-sensitive boosting for classification of imbalanced data. Pattern Recognition (40): 3358-3378

Yates W., Partridge D., Use of methodological diversity to improve neural network generalization,Neural Computing and Applications 4 (2) (1996) 114-128.

Yuan Y., Shaw M., Induction of fuzzy decision trees, Fuzzy Sets and Systems 69(1995):125-139.

Zadrozny B. and Elkan C. (2001): *Learning and Making Decisions When Costs and Probabilities are Both Unknown.* In Proceedings of the Seventh International Conference on Knowledge Discovery and Data Mining (KDD'01).

Zahn, C. T., Graph-theoretical methods for detecting and describing gestalt clusters. IEEE trans. Comput. C-20 (Apr.), 68-86, 1971.

Zaki, M. J., Ho C. T., Eds., Large- Scale Parallel Data Mining. New York: Springer- Verlag, 2000.

Zaki, M. J., Ho C. T., and Agrawal, R., Scalable parallel classification for data mining on shared- memory multiprocessors, in Proc. IEEE Int. Conf. Data Eng., Sydney, Australia, WKDD99, pp. 198– 205, 1999.

Zantema, H., and Bodlaender H. L., Finding Small Equivalent Decision Trees is Hard, International Journal of Foundations of Computer Science, 11(2):343-354, 2000.

Zeira, G., Maimon, O., Last, M. and Rokach, L,, Change detection in classification models of data mining, Data Mining in Time Series Databases. World Scientific Publishing, 2003.

Zenobi, G., and Cunningham, P. Using diversity in preparing ensembles of classifiers based on different feature subsets to minimize generalization error. In Proceedings of the European Conference on Machine Learning, 2001.

Zhang, C.X., Zhang, J.S., 2008. A local boosting algorithm for solving classification problems. Comput. Stat. Data Anal. 52 (4), 1928-1941.

Zhang, A., Wu, Z.L., Li, C.H., Fang, K.T., (2003), On hadamard-type output coding in multiclass learning. In: Proceedings of IDEAL. Volume 2690 of Lecture Notes in Computer Science., Springer-Verlag 397–404.

Zhang, C.X., Zhang, J.S. (2008), RotBoost: A technique for combining Rotation Forest and AdaBoost, Pattern Recognition Letters, Volume 29, pages 1524-1536.

Zhang, C.X., Zhang, J.S., Zhang G. Y., Using Boosting to prune Double-Bagging ensembles. Computational Statistics and Data Analysis, 53(4):1218-1231

Zhang, C.X., Zhang, J.S., Zhang G. Y., An efficient modified boosting method for solving classification problems , Journal of Computational and Applied Mathematics 214 (2008) 381 - 392

Zhou Z., Chen C., Hybrid decision tree, Knowledge-Based Systems 15, 515-528, 2002.

Zhou Z., Jiang Y., NeC4.5: Neural Ensemble Based C4.5, IEEE Transactions on Knowledge and Data Engineering, vol. 16, no. 6, pp. 770-773, Jun., 2004.

Zhou Z. H., and Tang, W., Selective Ensemble of Decision Trees, in Guoyin Wang, Qing Liu, Yiyu Yao, Andrzej Skowron (Eds.): Rough Sets, Fuzzy Sets, Data Mining, and Granular Computing, 9th International Conference, RSFDGrC, Chongqing, China, Proceedings. Lecture Notes in Computer Science 2639, pp.476-483, 2003.

Zhou, Z. H., Wu J., Tang W., Ensembling neural networks: many could be better than all. Artificial Intelligence 137: 239-263, 2002.

Zhoua J., Pengb H., Suenc C., Data-driven decomposition formulti-class classification, Pattern Recognition 41: 67 - 76, 2008.

Zimmermann H. J., Fuzzy Set Theory and its Applications, Springer, 4th edition, 2005.

Zitzler, E., Laumanns, M., Thiele, L., (2002), SPEA2: Improving the strength pareto evolutionary algorithm. In: Evolutionary Methods for Design, Optimisation, and Control, CIMNE, Barcelona, Spain. 95–100.

Zitzler, E., Laumanns, M., Bleuler, S., (2004), A tutorial on evolutionary multiobjective optimization. In Gandibleux, X., Sevaux, M., Srensen, K., T'kindt, V., eds.: Metaheuristics for Multiobjective Optimisation. Volume 535 of Lecture Notes in Economics and Mathematical Systems., Springer-Verlag 3–37.

Zupan, B., Bohanec, M., Demsar J., and Bratko, I., Feature transformation by function decomposition, IEEE intelligent systems & their applications, 13: 38-43, 1998.

Index